AFRICA JUNCTIONS
CAPTURING THE CITY

Lard Buurman

In happy memory of **Leon Erasmus**

AFRICA JUNCTIONS
CAPTURING THE CITY

Lard Buurman

WITH ESSAYS BY

Chris Abani
N'Goné Fall
Chris Keulemans
Alexander Opper

HATJE
CANTZ

CASABLANCA

ALEXANDRIA

CAIRO

DAKAR

LAGOS

DOUALA

KAMPALA

NAIROBI

KINSHASA

DAR ES SALAAM

LUANDA

LUBUMBASHI

LUSAKA

JOHANNESBURG

CAPE TOWN

Contents

Reinventing Reality

N'GONÉ FALL

Lard Buurman studied photography in the Department of Photography at the Royal Academy in The Hague. As he wanted to capture and share the reality of life around him, he opted for documentary photography. During his years at the academy, he intuitively grasped that it was possible to change the credibility of an image that the public took to represent an irrefutable reality. This unquestioned assumption—that a documentary photograph is a truthful document, an incontestable proof—is, in his view, a testament to a lack of a critical sense, which casts doubt on the free will of the public. The importance of the gaze of this public has led him to conceive an approach in which pure documentary and the narrative poetry of a photograph are combined. His aim is to invent stories based on factual reality. It was his analysis of the work of the Canadian artist, Jeff Wall, who dramatizes images, which prompted this decision and showed him the path he was to follow.

A FRAGMENT OF TRUTH

Lard Buurman thus decides to use the technique of photomontage, in which he combines public spaces and passersby to reinvent reality. Public spaces serve as the backdrop in his images. It is the décor in his scenes that tell stories, which are in their appearance anecdotic. The characters that he selects and subsequently inserts into the photographs transform the space and conjure up events that might have taken place at another time. The passerby becomes an actor and the city the décor of a moment of life to be evoked through a single freeze-frame. This process of creating a document resembles the production of a film where he would be at once the scriptwriter, the casting director, the director, and the editor. Lard Buurman follows the logic of modified reality to the very limit. What one perceives is just a fragment of truth. The documentary photograph, absolute proof, becomes open to doubt. Just as there are as many versions of one event as there are witnesses, Buurman's images are similar in this respect, because they present a point of view that is open to interpretation. Taking an image now, or two seconds later is a choice. Choosing one character rather than another is also a choice. These deliberate choices made by the photographer are in themselves the affirmation of the fragility of a truth that questions the reliability of documentary photography.

Buurman's photographs exude a strange, undefined sensation. All the characters have a perfect posture that reflects the setting where they are placed. His artistic approach, which consists of portraying reinvented reality, recalls that of the studio photographers of the nineteenth century and also the great masters of European painting: the setting and figures illustrate everyday life. And as he always works in the open air, all his urban scenes are idealized moments of the public space.

EUROPE AND CHINA

Buurman's own continent provides the framework for his first works, the series *Taking Time* (1999 / 2002), where he photographs medium-sized cities such as Stuttgart, Dublin, and Lisbon, avoiding the trap of producing the umpteenth vision of major capitals. In this early work he is able to distance himself, and is especially able to look with new eyes at urban spaces—the soul and vibrations of a city. Later, in Amsterdam, he concentrates on images of construction sites. *For the Time Being* (2001 / 2004) depicts the rigor and discipline of Dutch urban spaces where even a building site is a "clean," un-chaotic space uncluttered by litter. And the workers, industrious and properly dressed, reinforce the myth that in the cities of the Western world, order and a sense of well-being reign supreme—a kind of fable of the radiant city that all industrialized and civilized cities must express. He appropriates this mythical notion of the ideal city, somewhere between political propaganda and popular fantasy, and undertakes the task of visiting developing countries to look for contrasts and for lands that are not artificially designed and have more personality. During his visit to China he realizes that the distinction between city and countryside is not as obvious as in the West. Here, the rural culture is a component of the urban lifestyle. The series *Peoples Republic of China* (2004 / 2007) depicts mushrooming cities, population displacement, and urban tension, which represent only the visible face of profound economic and social changes in the country. Here, urban landscapes are changing more quickly than anywhere else. And, in one way or another, they are the ones imposing a new way of life on the people. In this series, the image of China with an age-old culture, with a disciplined population working with pride for the public good, is attacked. Lard Buurman's Chinese cities seem to make us believe that the Maoist doctrine is dead—sacrificed on the altar of progress in the name of a modernization in which nothing is equitable. The African continent is the next step of his journey in developing countries. His long-term plan is to create a corpus of photographs depicting the inhabited zones of the planet, and based on contrasts. A structured and strictly regulated urban West, as opposed to a seemingly disorganized Third World where rural and urban cultures naturally cohabit and where the informal is the obvious *mode de vie*.

AFRICA

In *African Junctions* (2008 / 2014), he tackles the African continent, visiting fourteen cities from North to South and East to West, passing through Central Africa. He immerses himself, among others, in Cairo and Casablanca, in Lusaka and Luanda, and Dakar and Kinshasa. In Zambia, the first stage of his African adventure, he realizes that an ever-increasing population is concentrated in the capital, that the capital is the cultural and economic center that is in fact forging the identity of the country, and that it is the urban space rather than the rural area that is the real mirror of society. In taking on Africa, Lard Buurman is aware that he might be perceived as yet another white person who turns up to photograph poverty and satisfy his fantasy of being an empathetic latter-day missionary. But Buurman is not a naïve man. His trips will give him the opportunity to take down various clichés diffused by Western media. Africa is not a vast territory composed of rural people attached to their ancestral traditions and impervious to the temptations of modernity. Africa is not a dying continent besieged by shantytowns, poverty, social tension, and permanent insecurity. Instead, Buurman illustrates that Africa, perceived as a continent with problems, is also a continent of enticing possibilities, in tune with the rest of the world.

This book is dedicated to Africa. There are fourteen cities in eighty images, which are not representations of a continent, nor illustrations of an Africa that discourages as much as it invites compassion. *Africa Junctions*, like the preceding series, is a suite of portrayals of urban space. And even if I am aware of the process of "constructing reality" used by Lard Buurman, these cities are not faithful enough to the memories that I have of them. Johannesburg, Cairo, and Lagos have a disarming urban density. Luanda, plunged into a frenzy of intoxicating reconstruction, changes in appearance every day. Dakar, Casablanca, and Cape Town exude a benevolent colonial nonchalance. Douala, Dar es Salaam, Kampala, and Nairobi are still struggling to charm me, while Lusaka is searching for its center and is coming to terms with its provinciality. And Kinshasa, faithful to its myth, evokes the sense of a joyous apocalypse. Well, that is not the case in Lard Buurman's photographs. By avoiding postcard images (souks, markets, and beach resorts), the residential areas of the elite, as well as emblematic monuments to the glory of independence and people fallen at the barricades, he reveals popular average areas with their improbable infrastructure, and a reserved middle class strolling through the city's main thoroughfares. Disappointment? Frustration? If African cities, as viewed by Lard Buurman, match neither the vision nor the sensations of the intracontinental globe trotter that I am, they do indeed successfully convey what they are: Third World cities with their invisible elite, who, like a cohort of ghosts, circulate only in their air-conditioned SUVs or limousines; areas where, come what may, the public space belongs to everyone, traders, artisans, idle adolescents, and beggars. Lard Buurman's African cities faithfully reflect the idea that one has of a Third World city, at the other end of the world: organized chaos, with its own principles and urban code, a voracious energy composed of noise, color, and bursts of laughter at the absurdity of the world. Whether it's a business district or a garage open to the sky, all the characters display the same sobriety. Neither poor souls in search of civilization or salvation, nor victims of selfish Western globalization, they have the same expressions and attitudes as the people photographed in China or in Europe. They are always anonymous silhouettes moving through life in the certitude that the best is still to come.

One has to distance oneself to look at any context in a new way and analyze a situation objectively. Probably because I've walked too often through these cities, I've ended up no longer able to see them and am rather content with bitter or sweet sensations linked to personal encounters, the architectural heritage, or the climate. Lard Buurman plunges me into a reality that is only partially mine. A raw reality stripped of feverish false poetry, melancholy, and artificial nostalgia. This reality, reinvented through the montage of simple scenes of everyday life, comforts and reassures me in another truly real certainty: in the Third World, life happens on the streets, and the city reinvents itself incessantly through the grace of its inhabitants, be they underdogs, working class, or the pedantic rich.
Africa Junctions **isn't a compilation of images representing Africa. This book is a series of modified truths.** ***Africa Junctions*** **is just a stage in that odyssey started in 1999 when Lard Buurman decided to challenge our perceptions of reality and the passing of time through the use of urban space by different cultures of the planet.**

JUNCTION #1

CITY OF STRUCTURES

What Buurman's images do is to shift our focus from a dumbed-down desire for Western-premised typological spatial foci (such as "public" squares, "public" parks, and so forth) —as easily measurable and categorical urban building blocks—to careful observations centered on "people as infrastructure."*

* Alexander Opper, "Recognizing and Revealing the Public," pp. 206–212

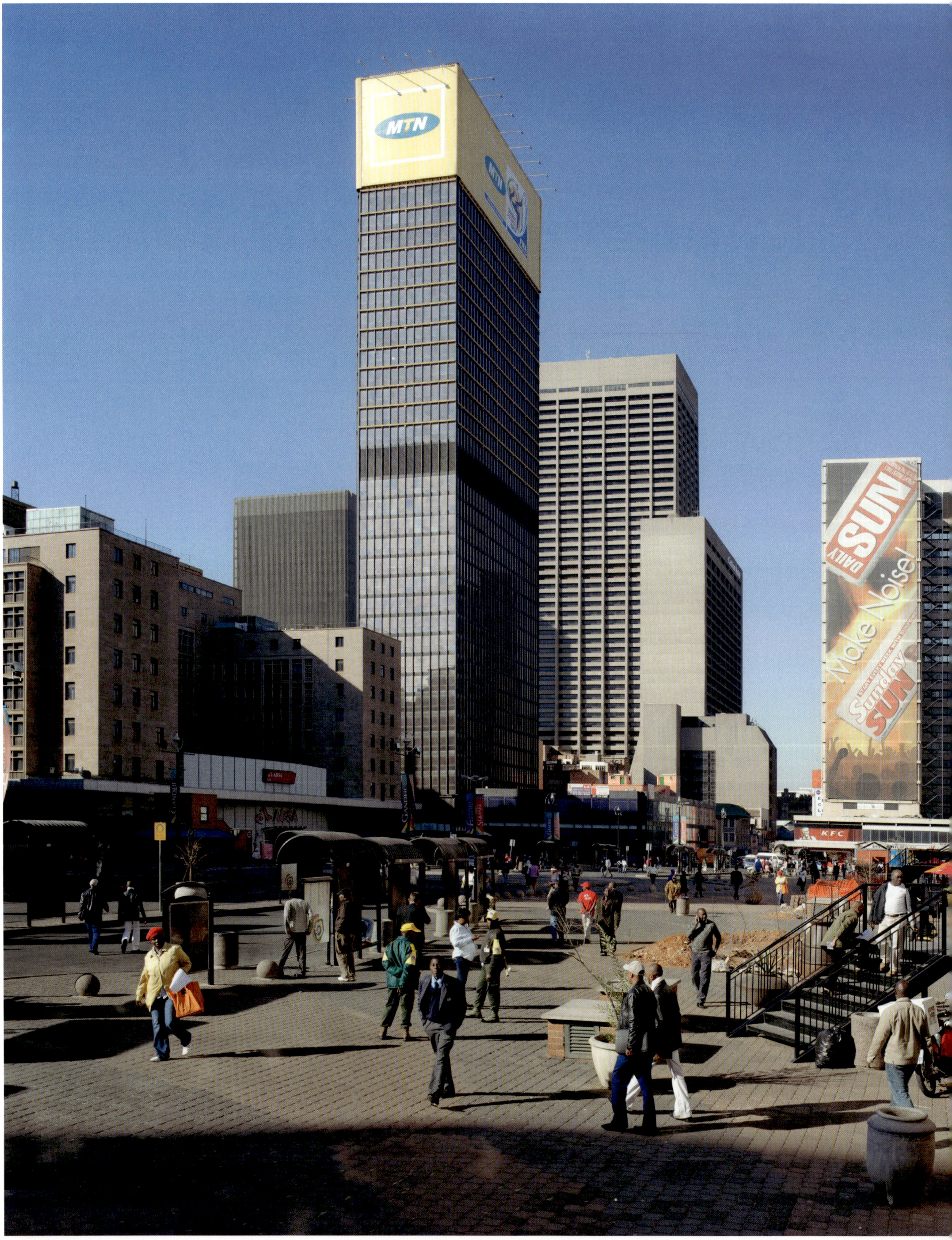
MTN
DAILY SUN
Make Noise!
Sunday SUN
KFC

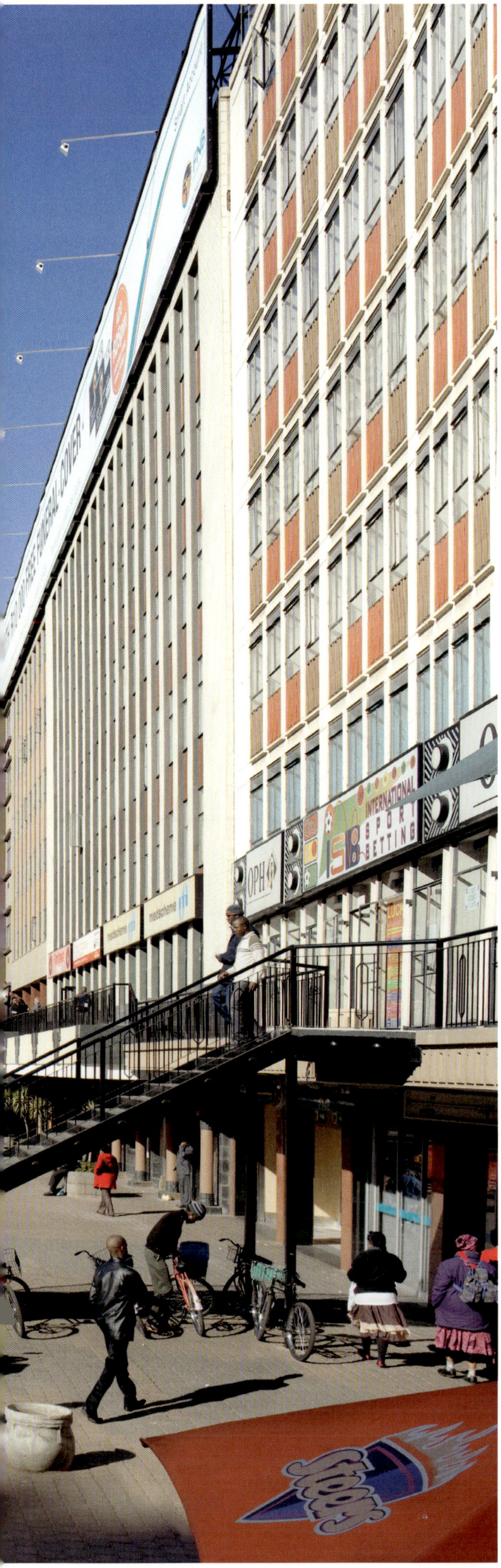

GANDHI SQUARE
JOHANNESBURG

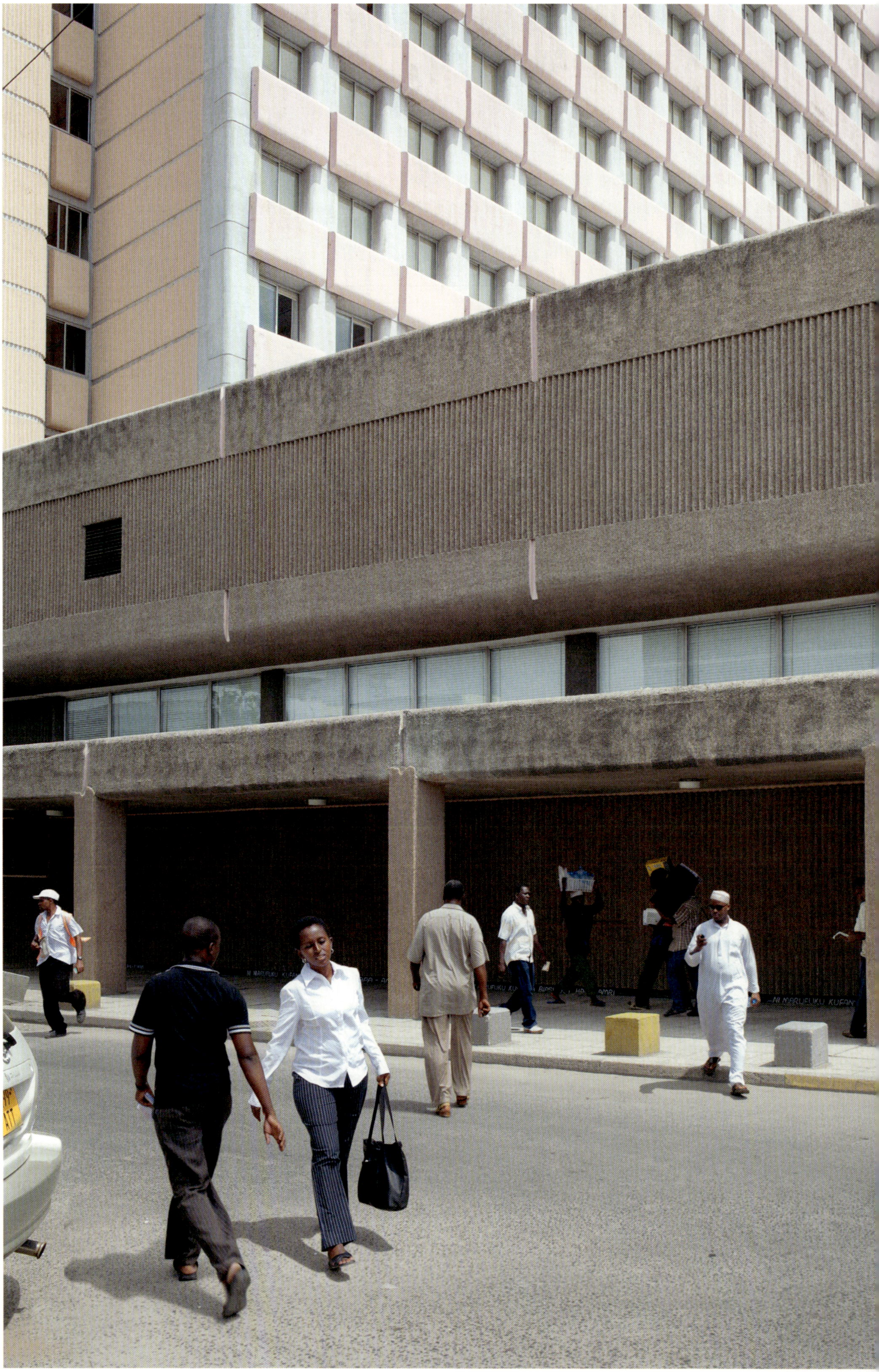

BRIDGE STREET
DAR ES SALAAM

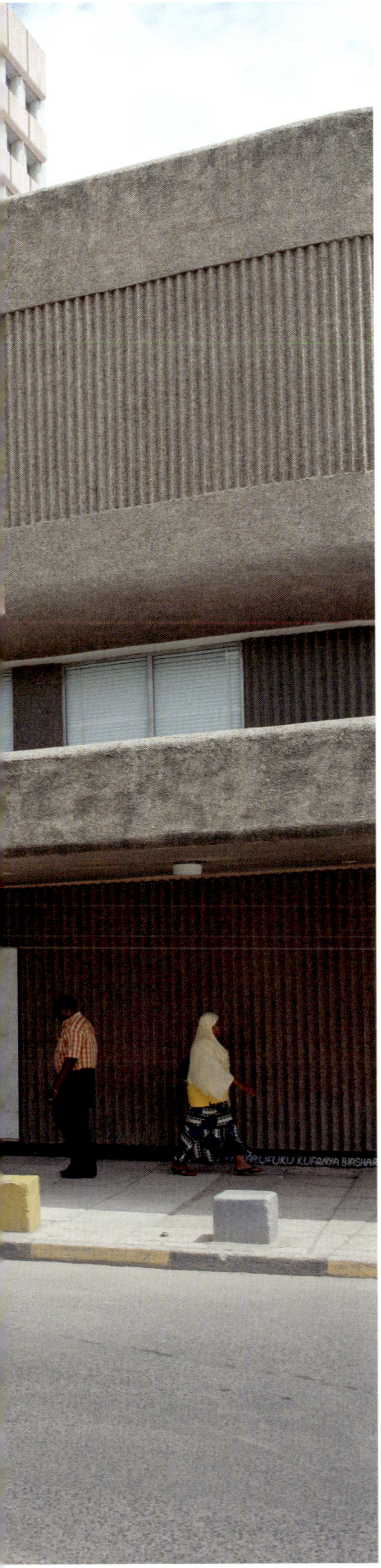
RUFUKU KUFANYA BIASHAR

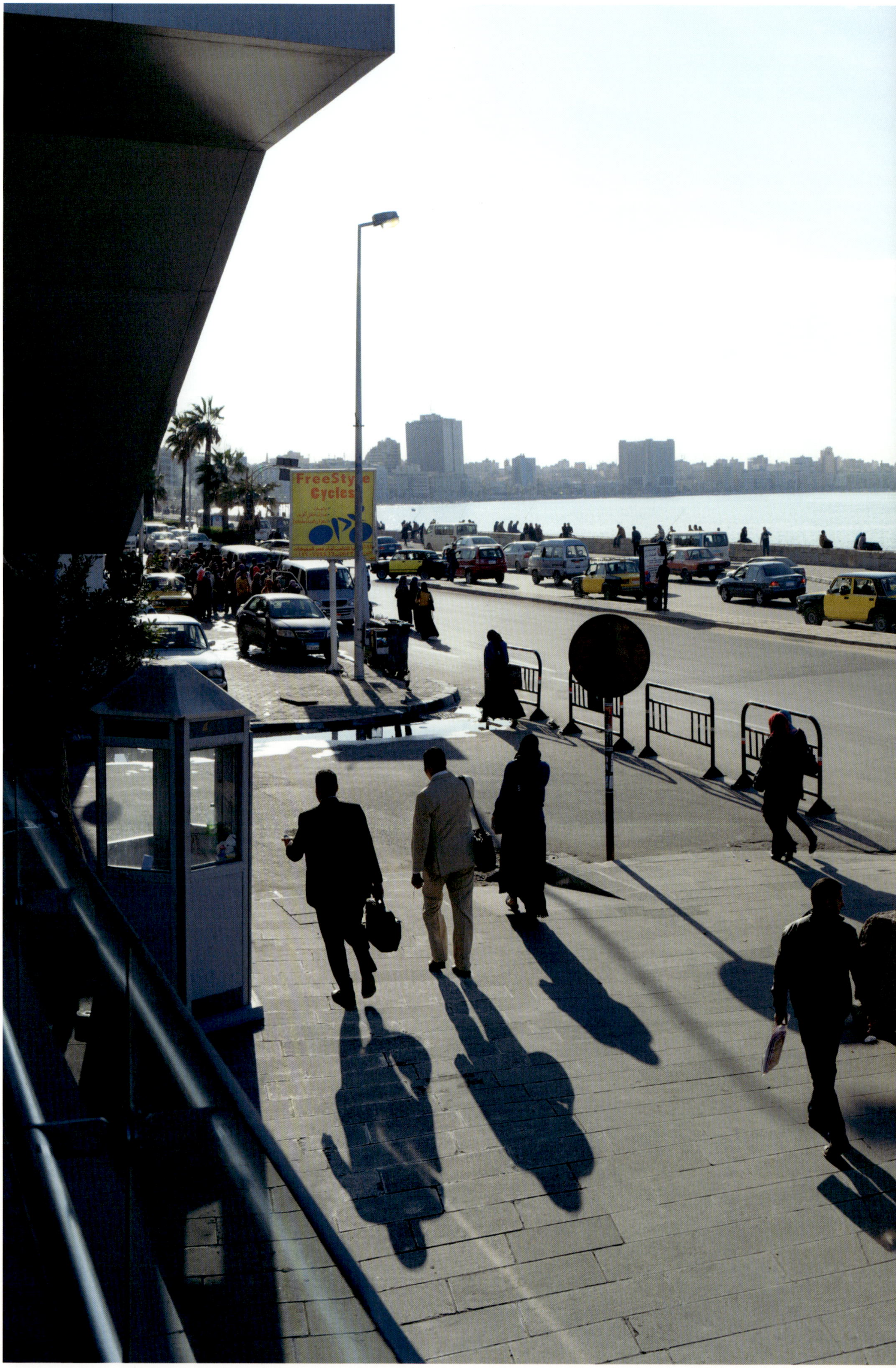
Cycles

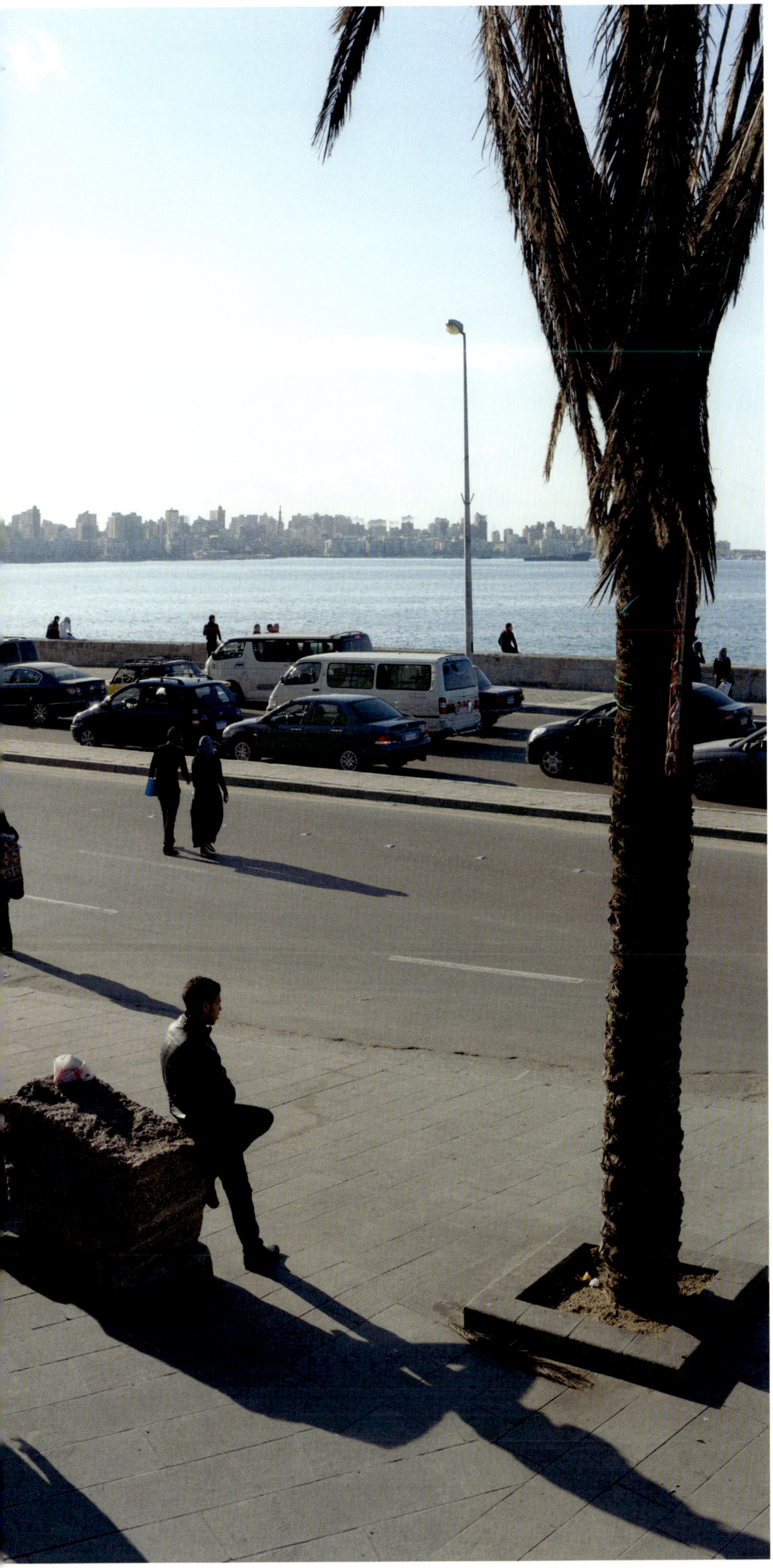

26 JULY EL-GAISH ROAD
ALEXANDRIA

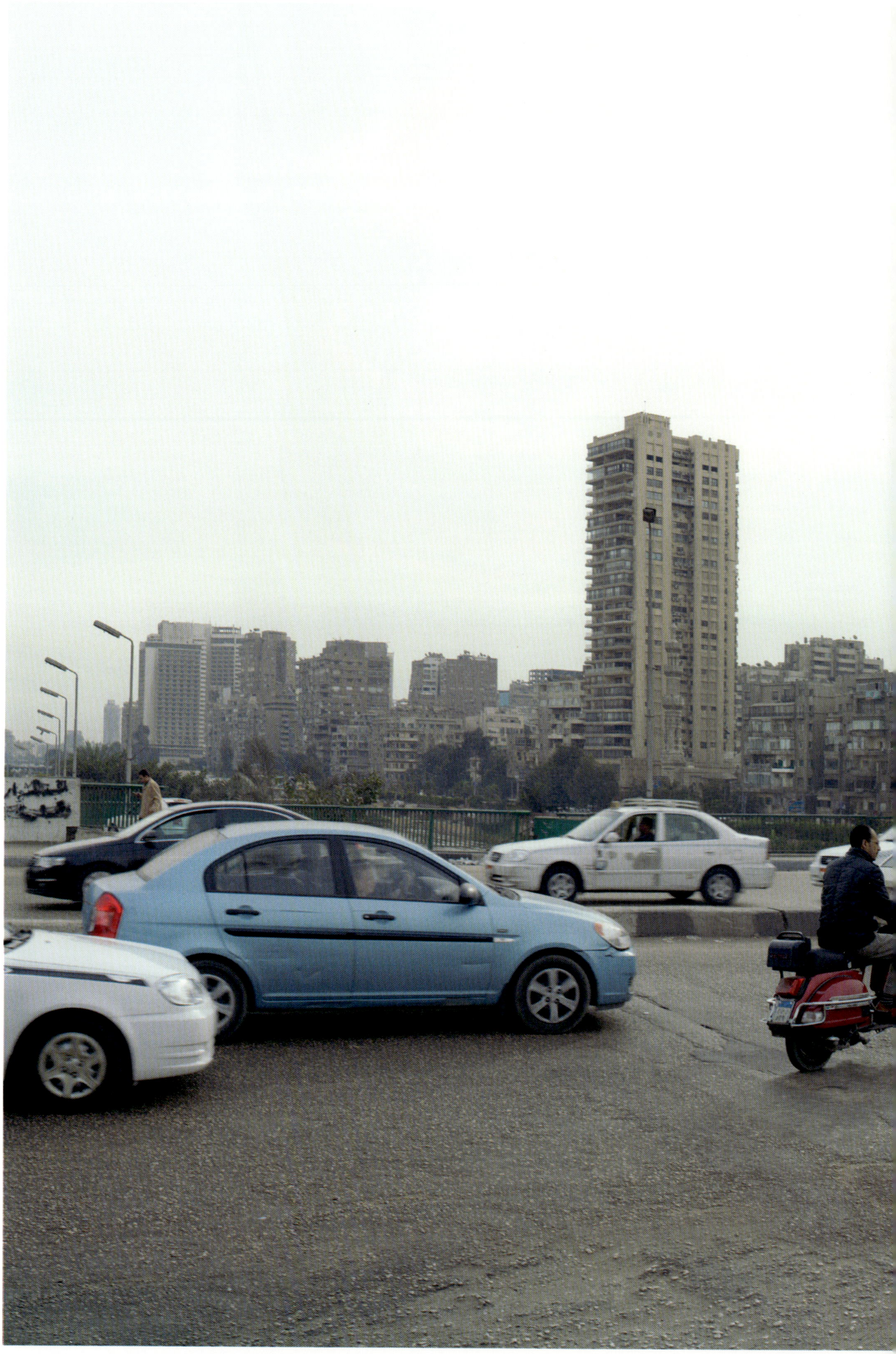

6TH OCTOBER BRIDGE
CAIRO

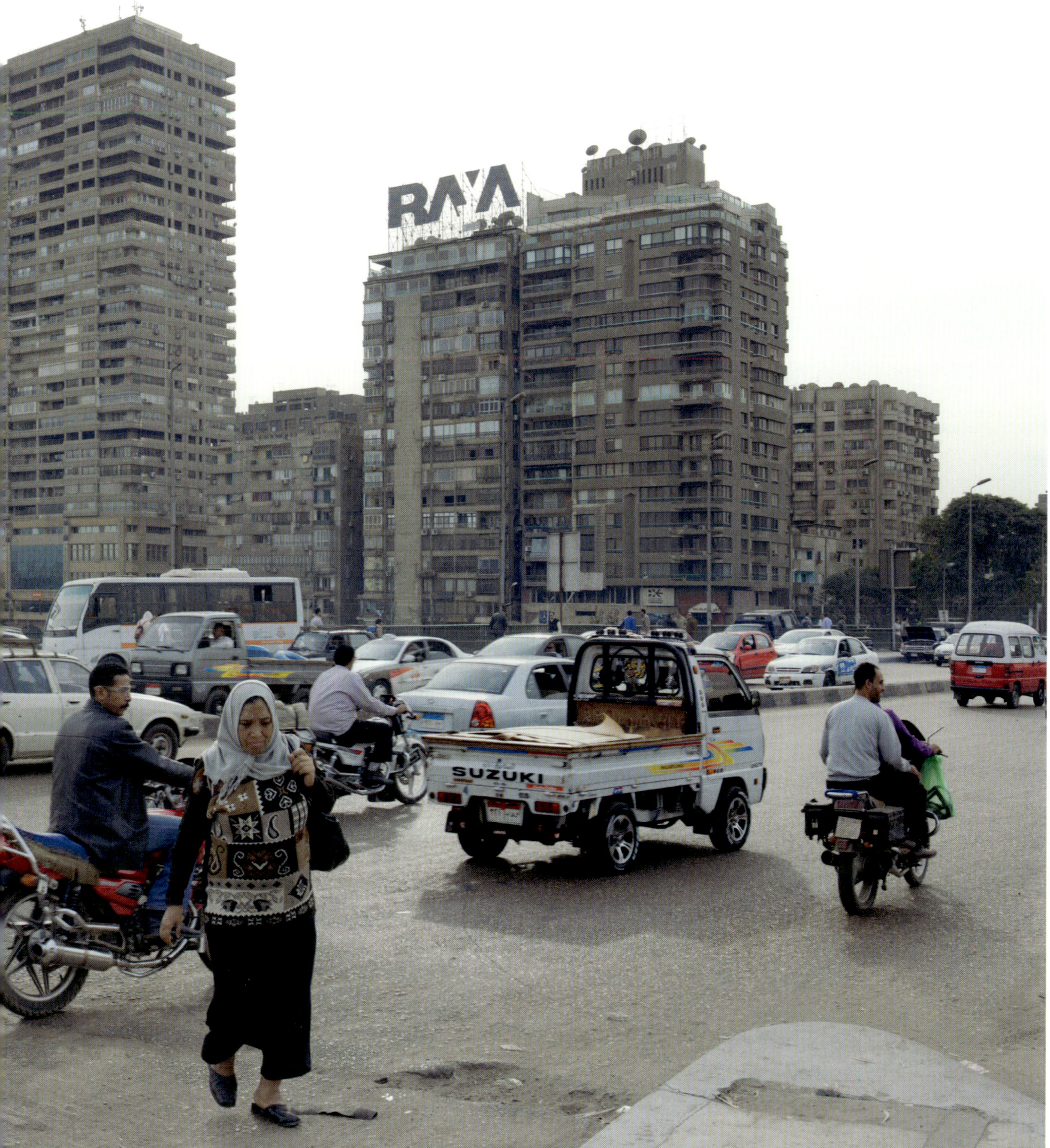
RAYA
SUZUKI

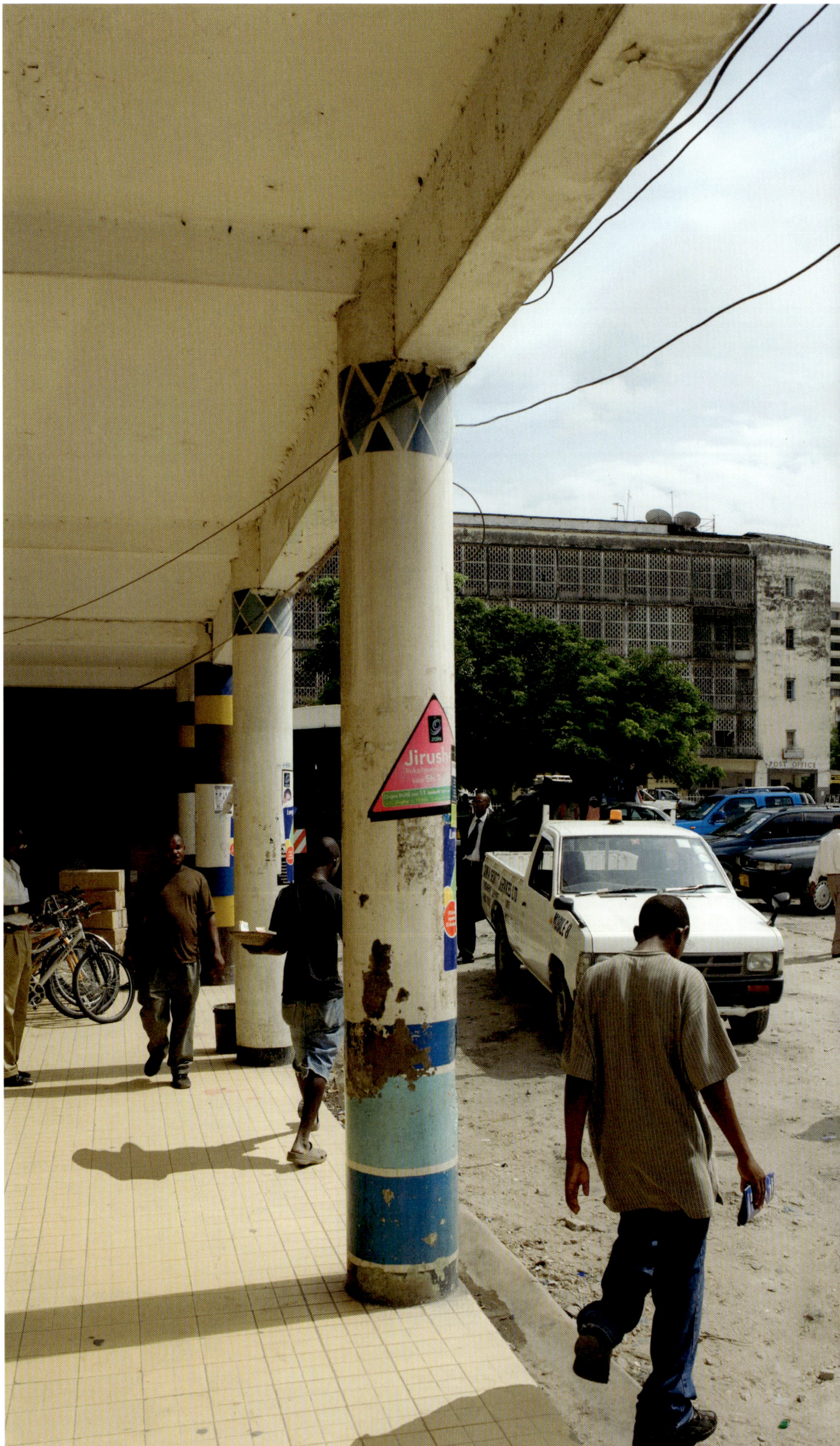
Jirush
POST OFFICE

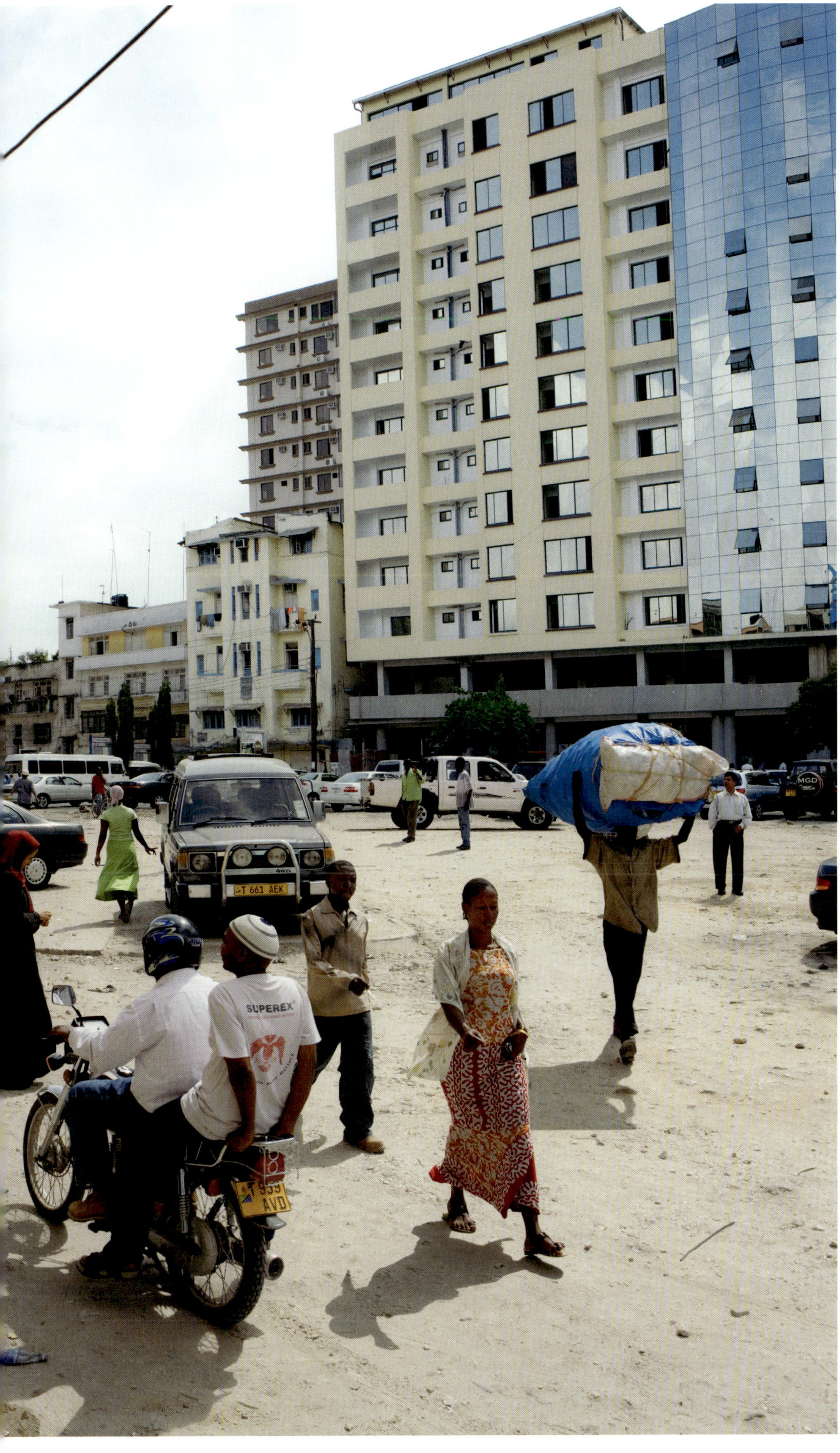
SUPEREX
T 661 AEK
T 959
AVD

CAMPBELL STREET
LAGOS

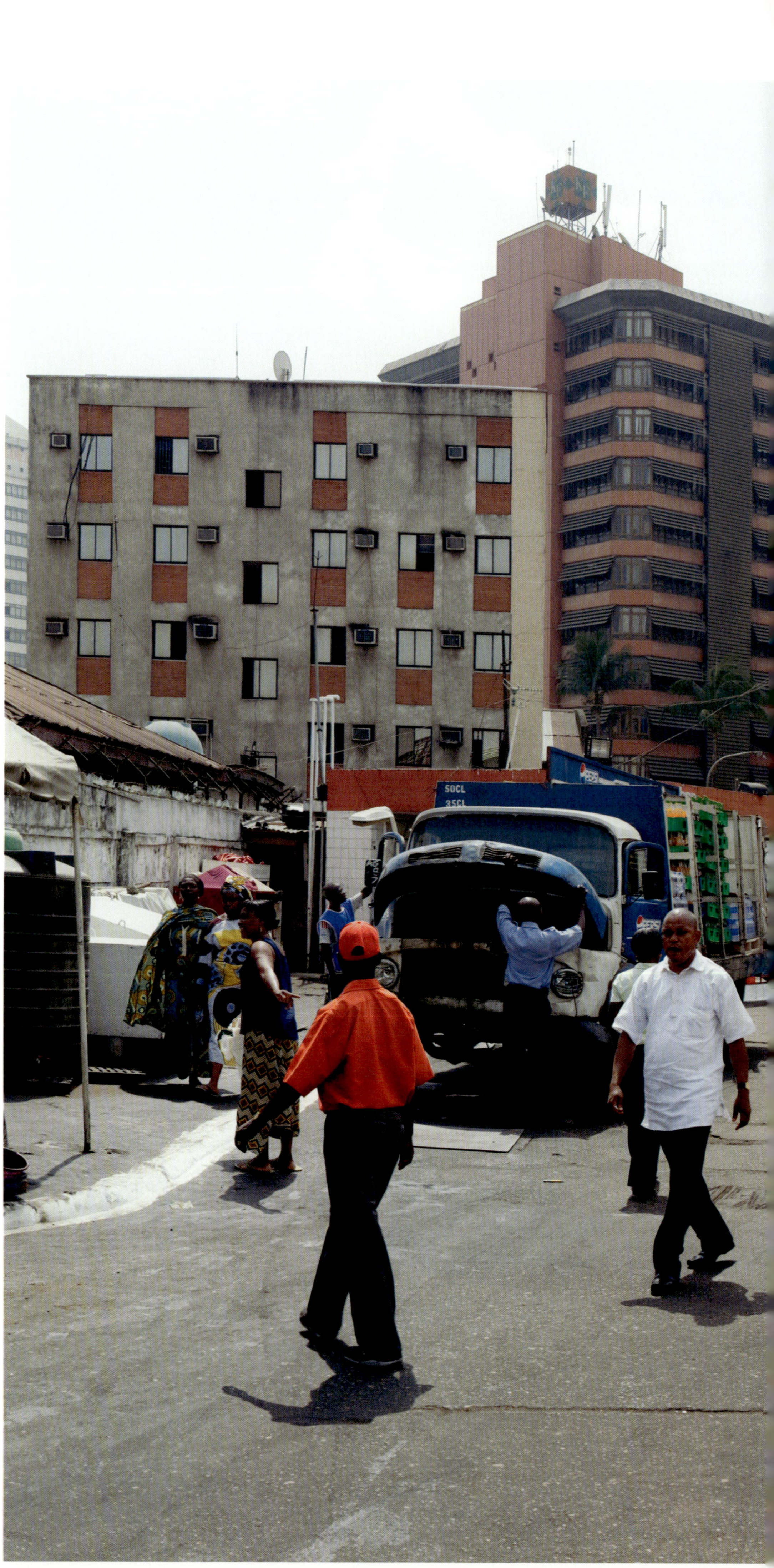

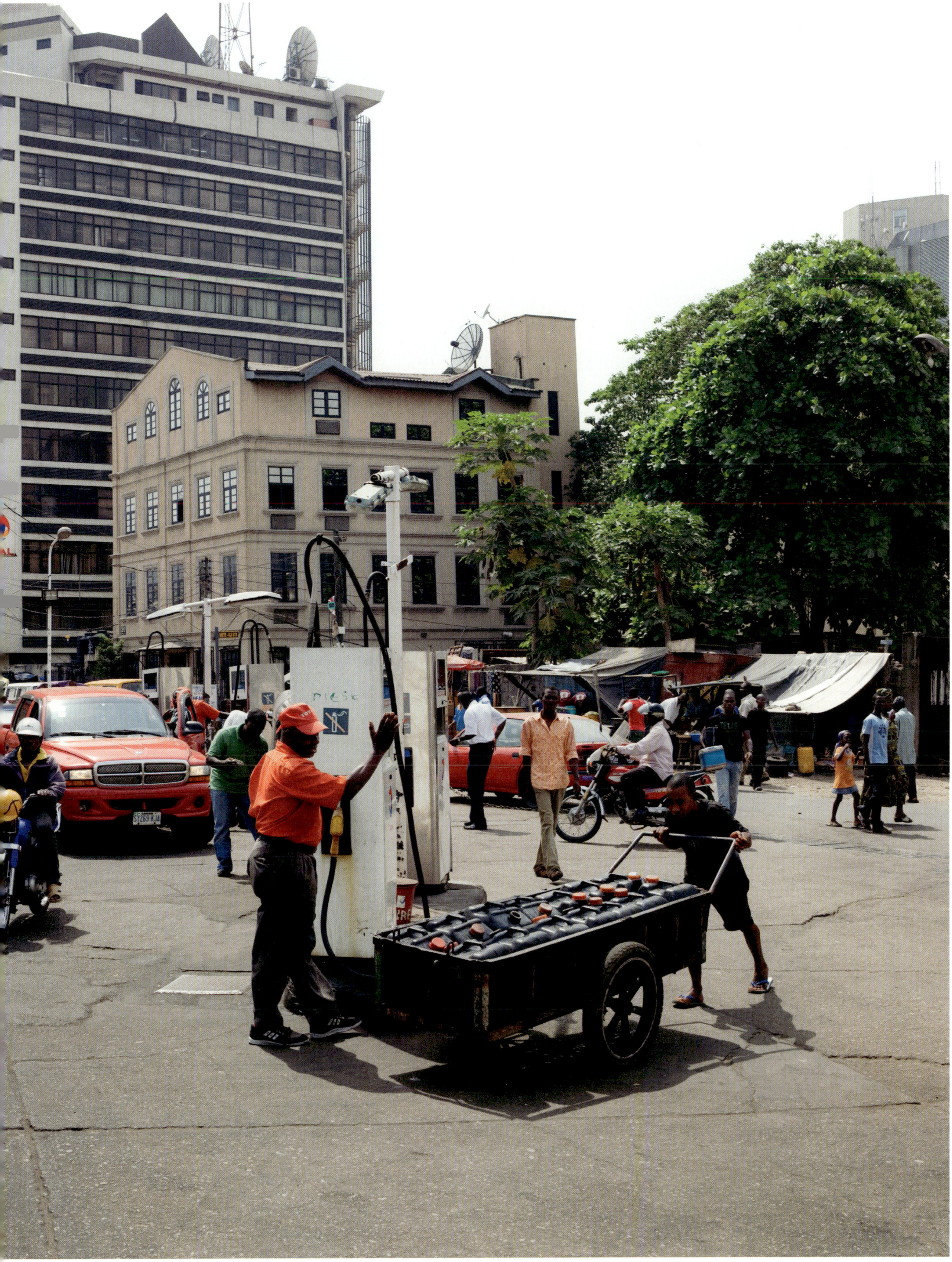

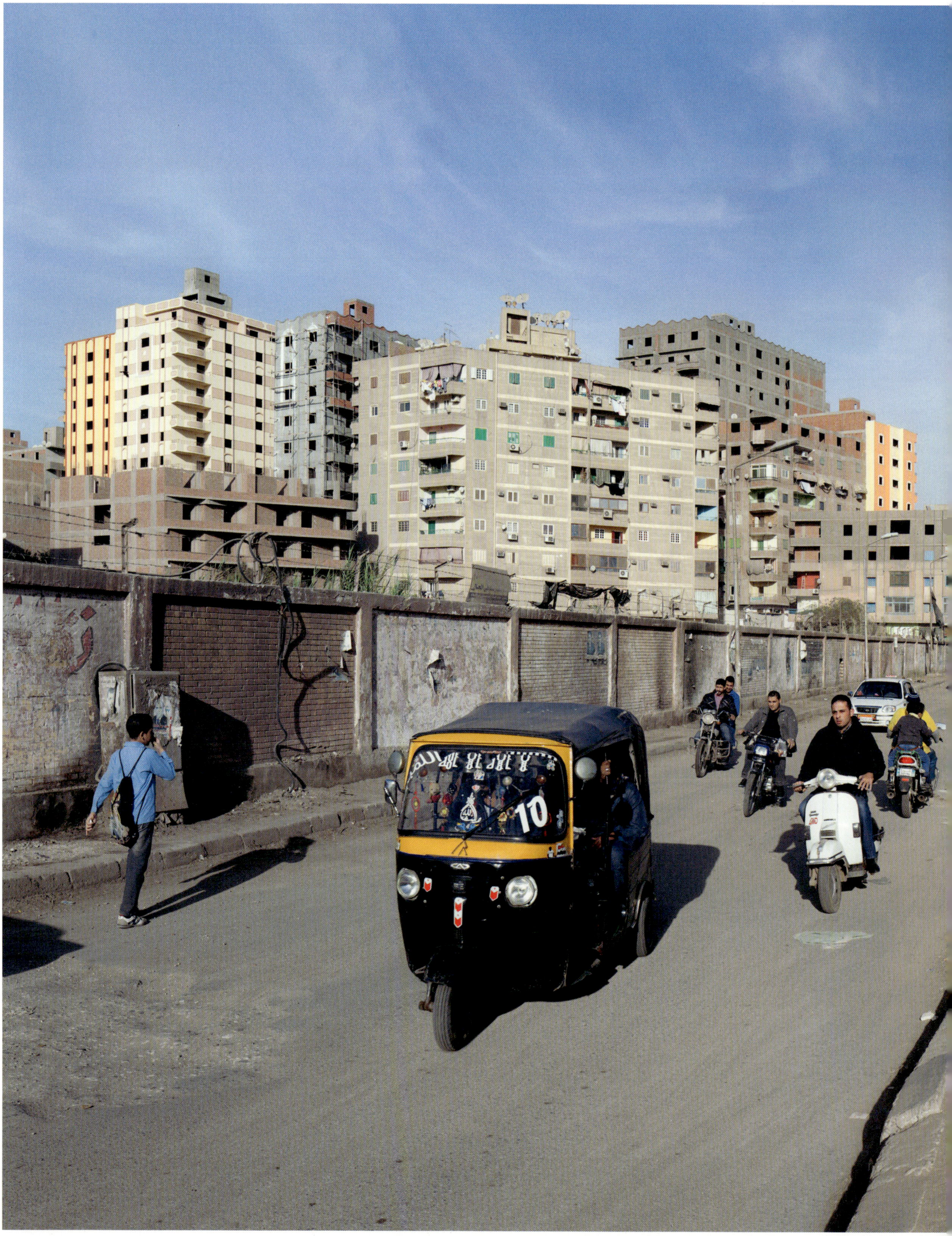
10

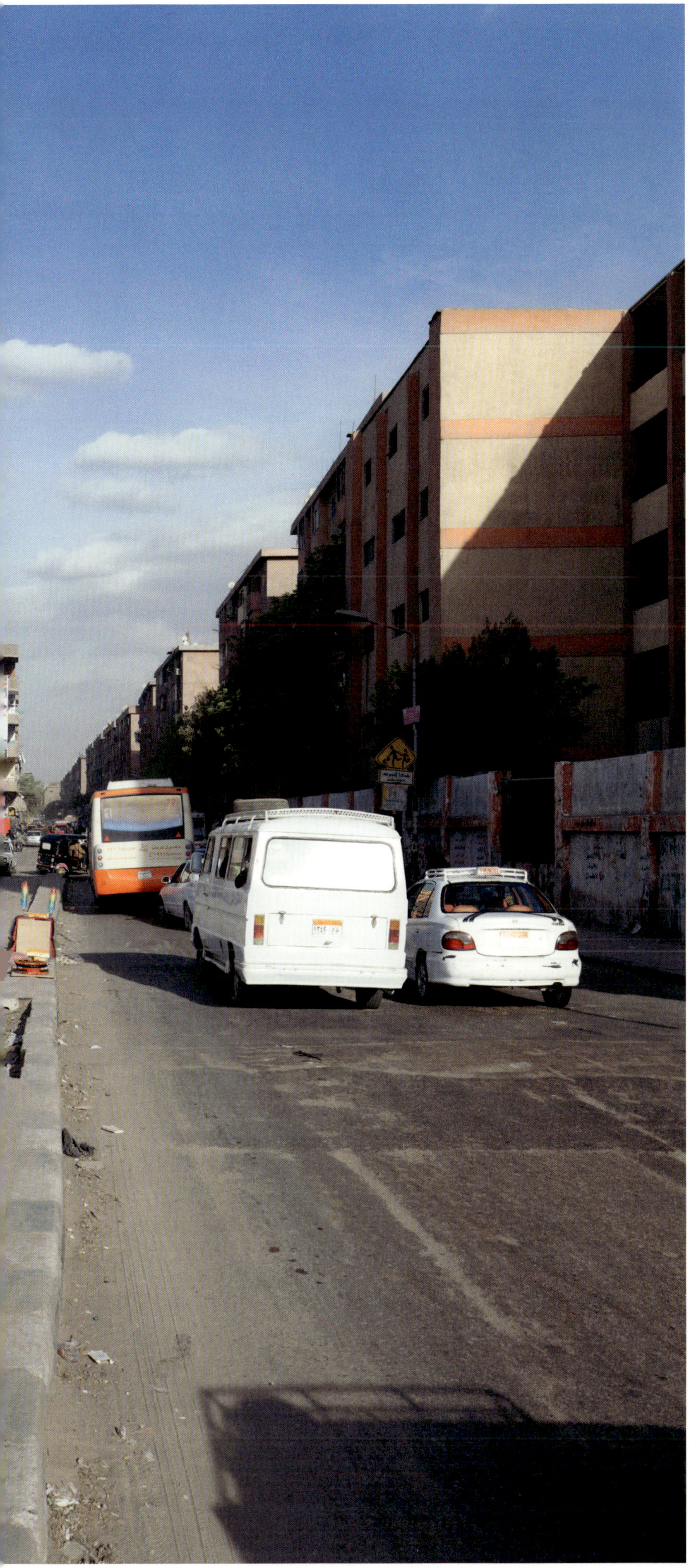

MATAR STREET
CAIRO

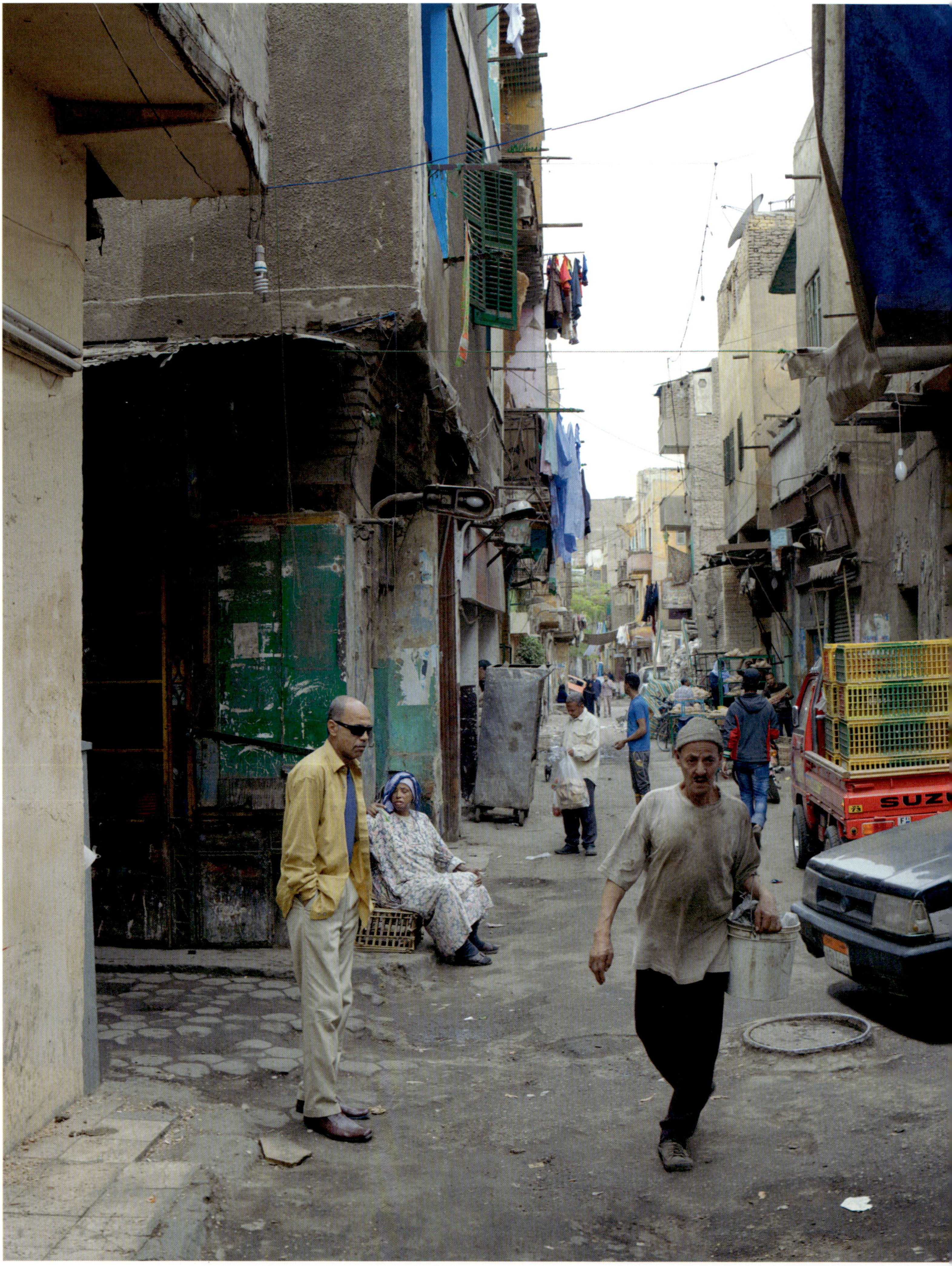
SUZU

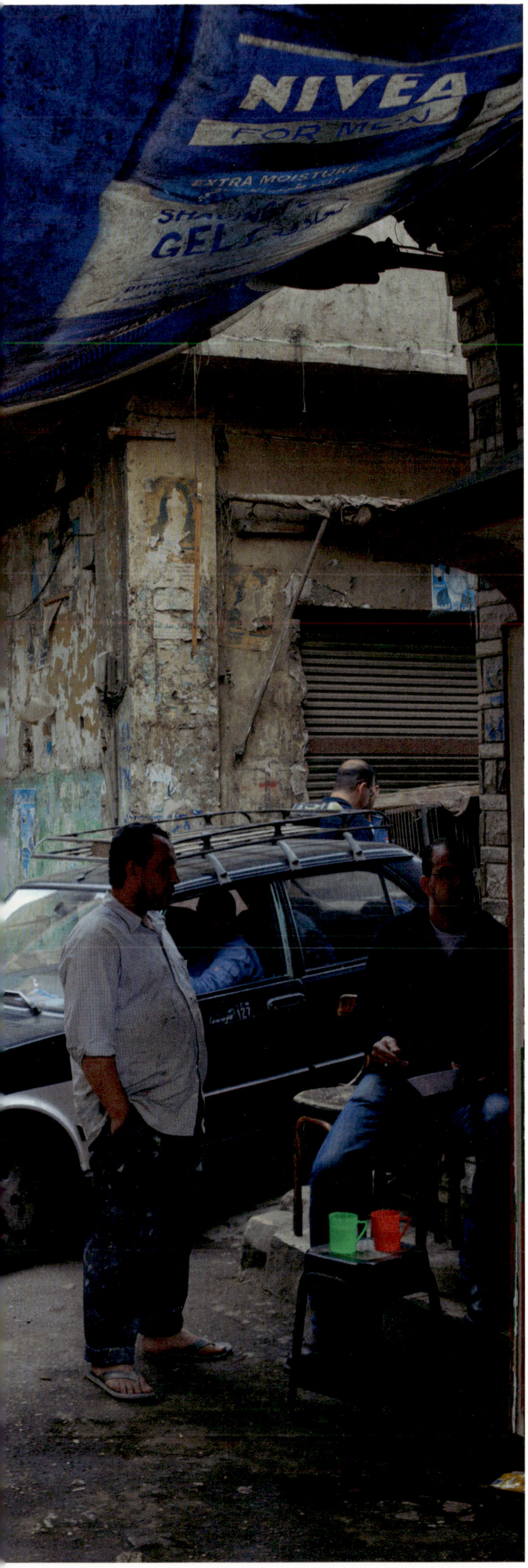

ABU TALEB, BULAQ
CAIRO

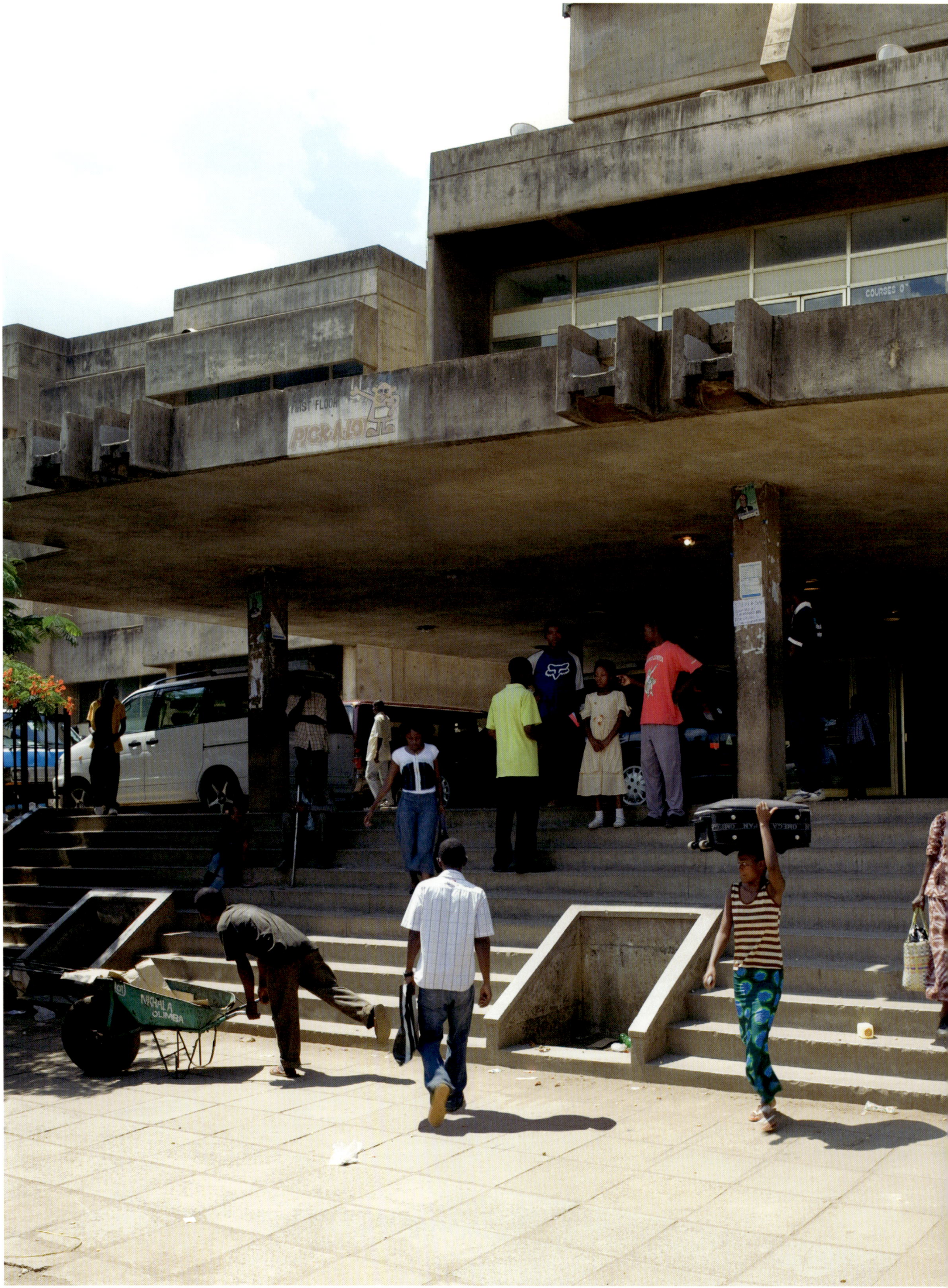
FIRST FLOOR
NKHALA
OLIMBA

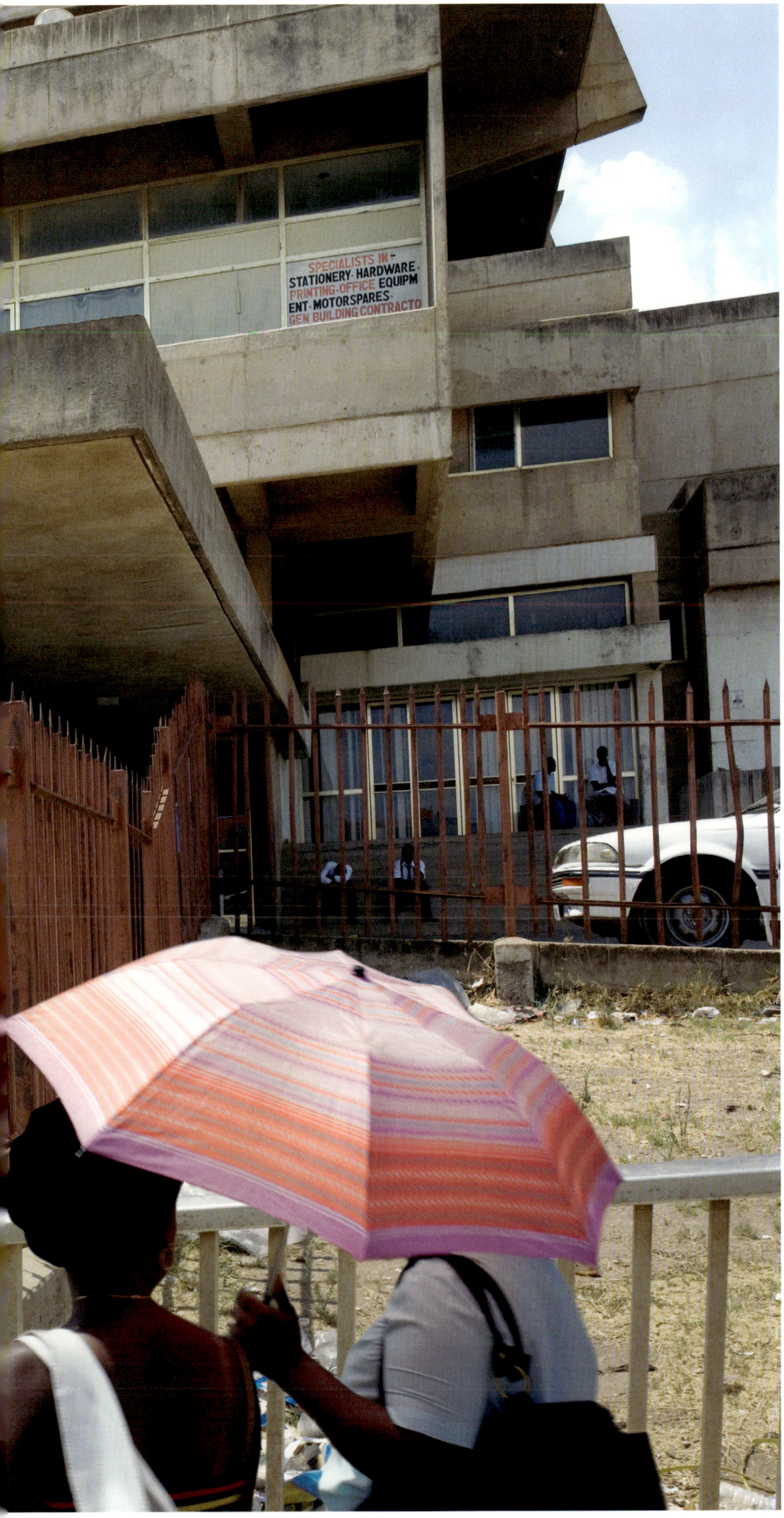

FINDECO ENTRANCE
LUSAKA

APOGBON
LAGOS

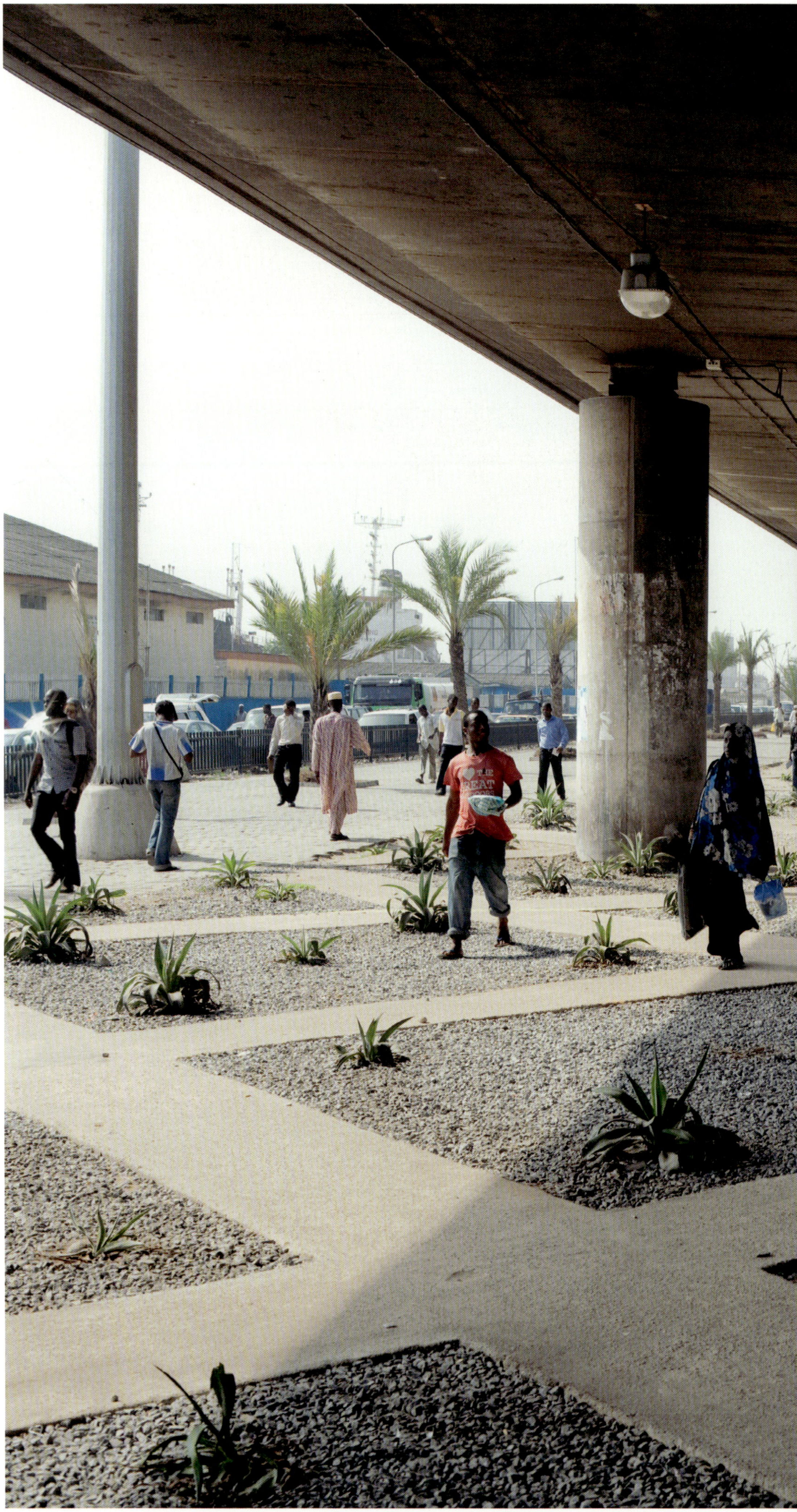

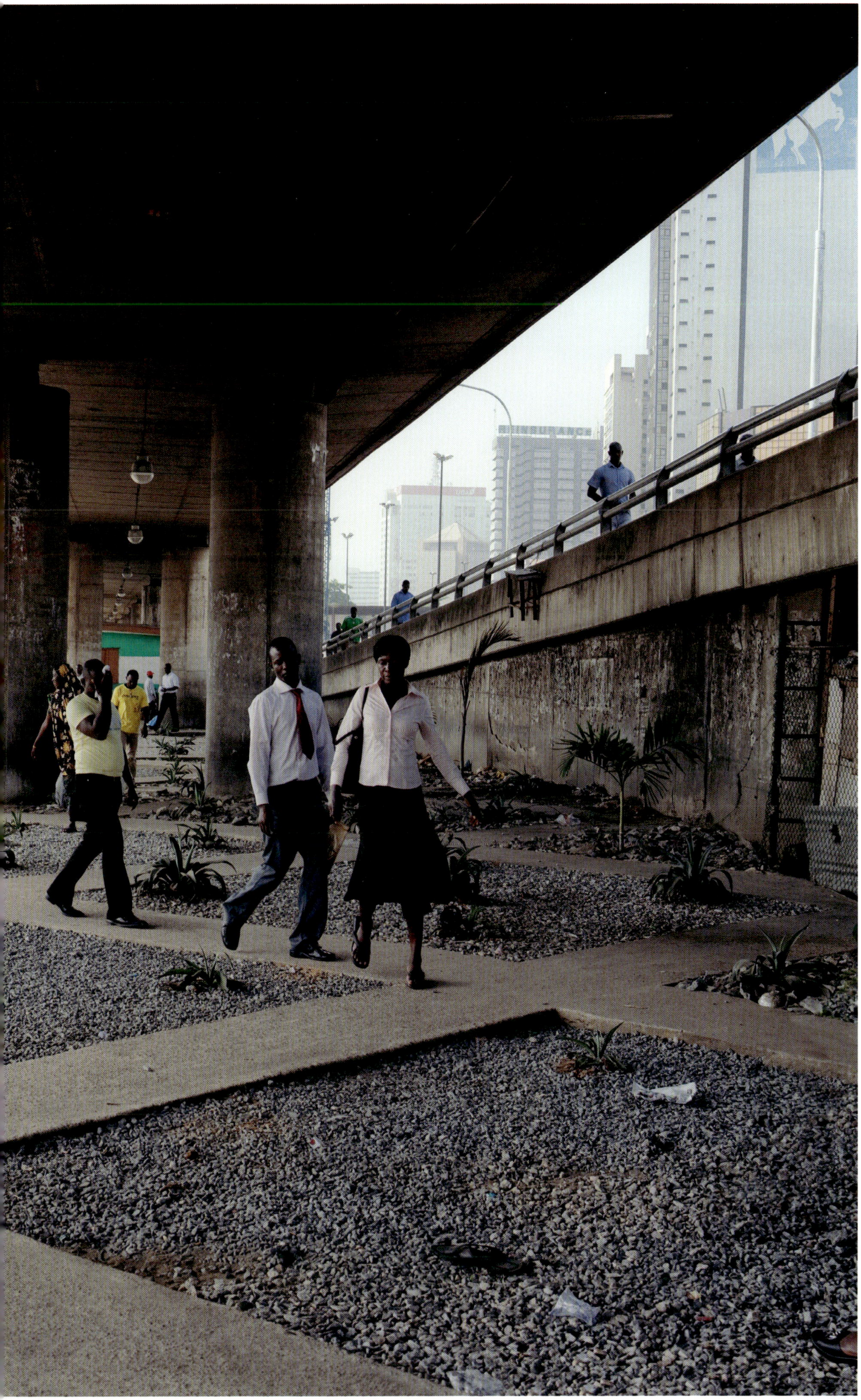
INSURANCE

BOULEVARD AL JOULAN
CASABLANCA

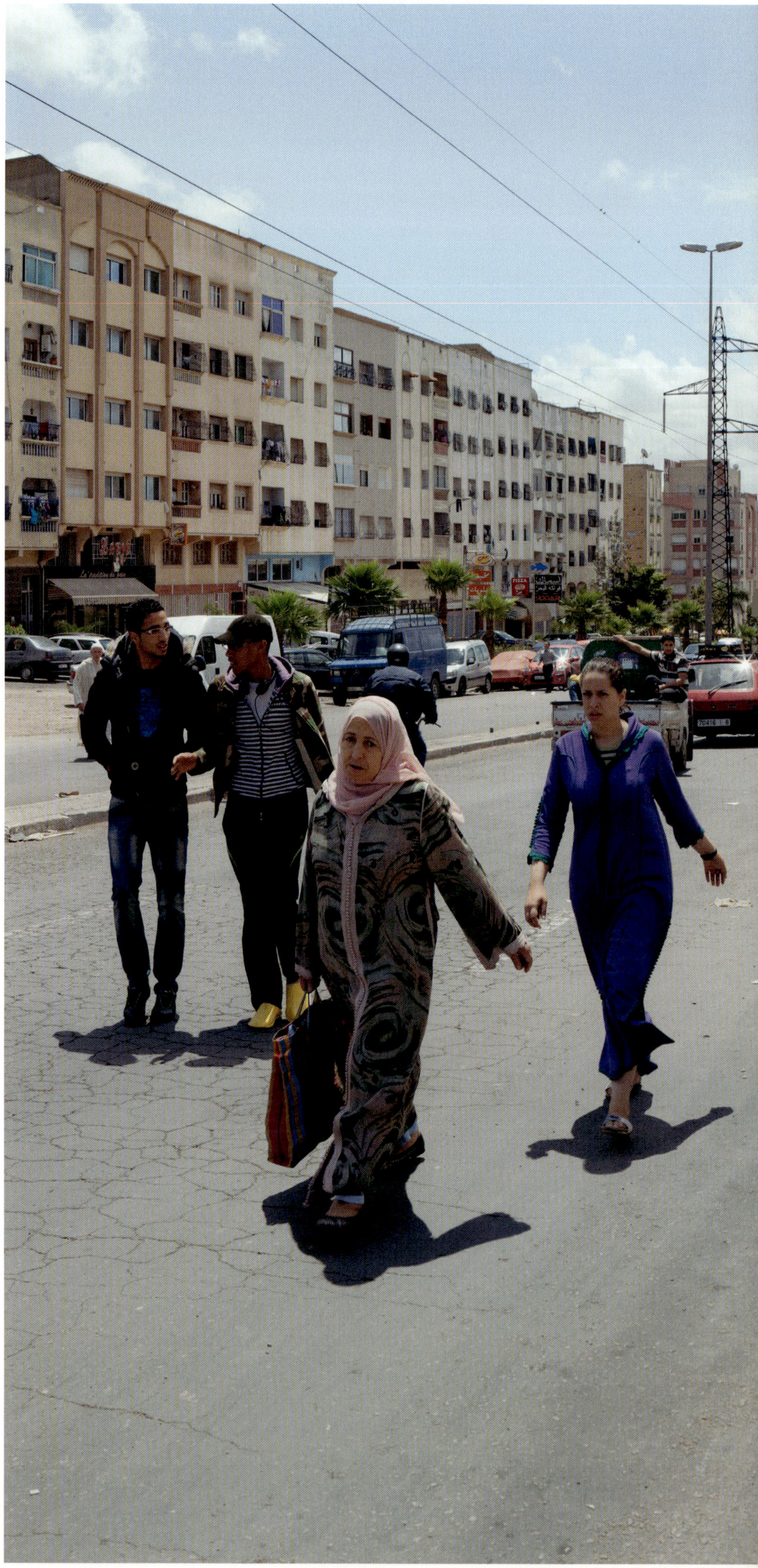

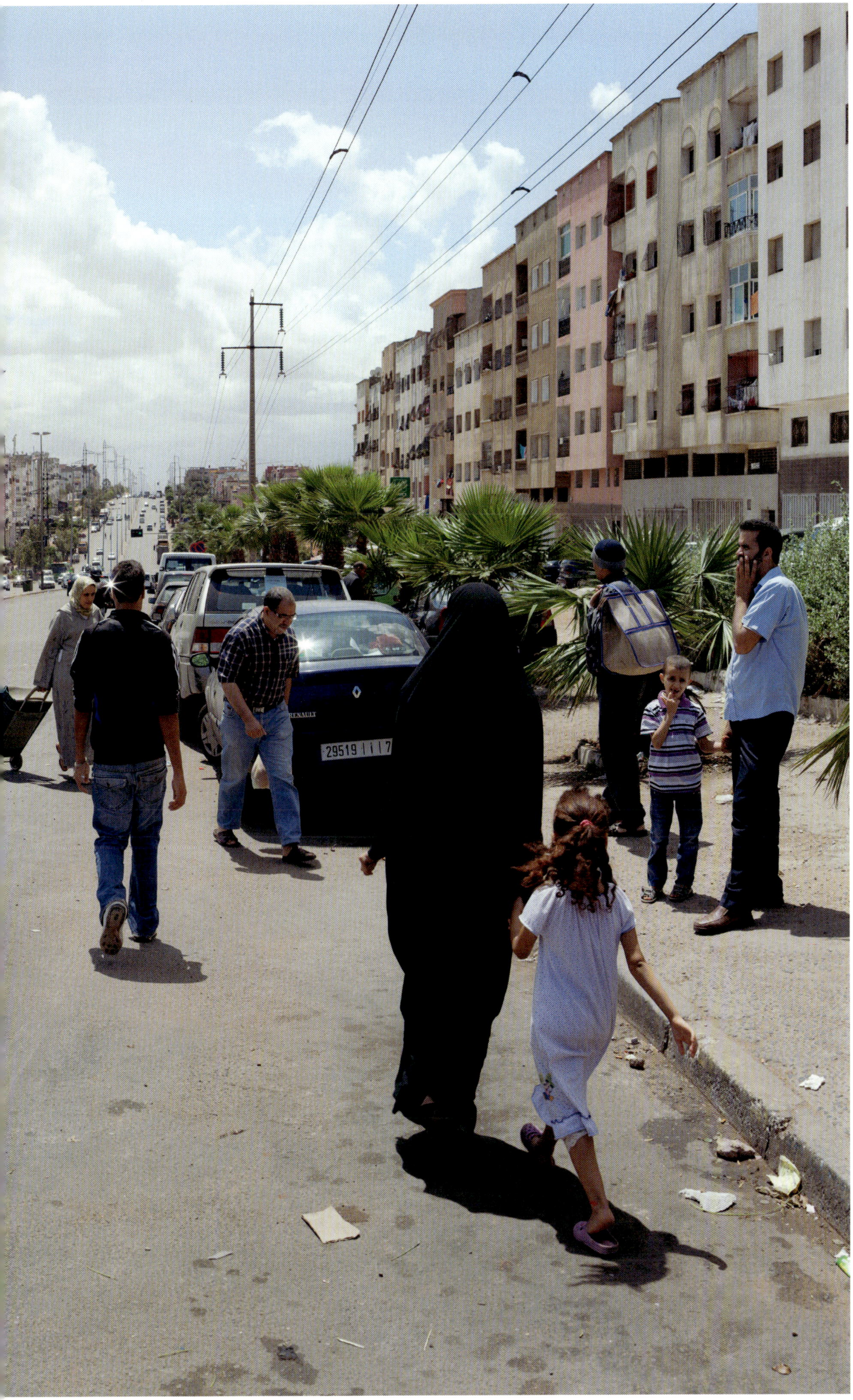
RENAULT

AVENIDA DE PORTUGAL
LUANDA

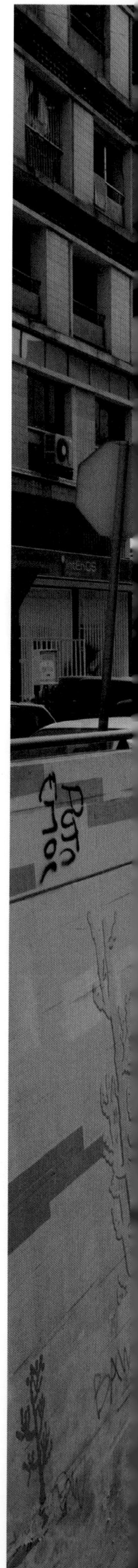

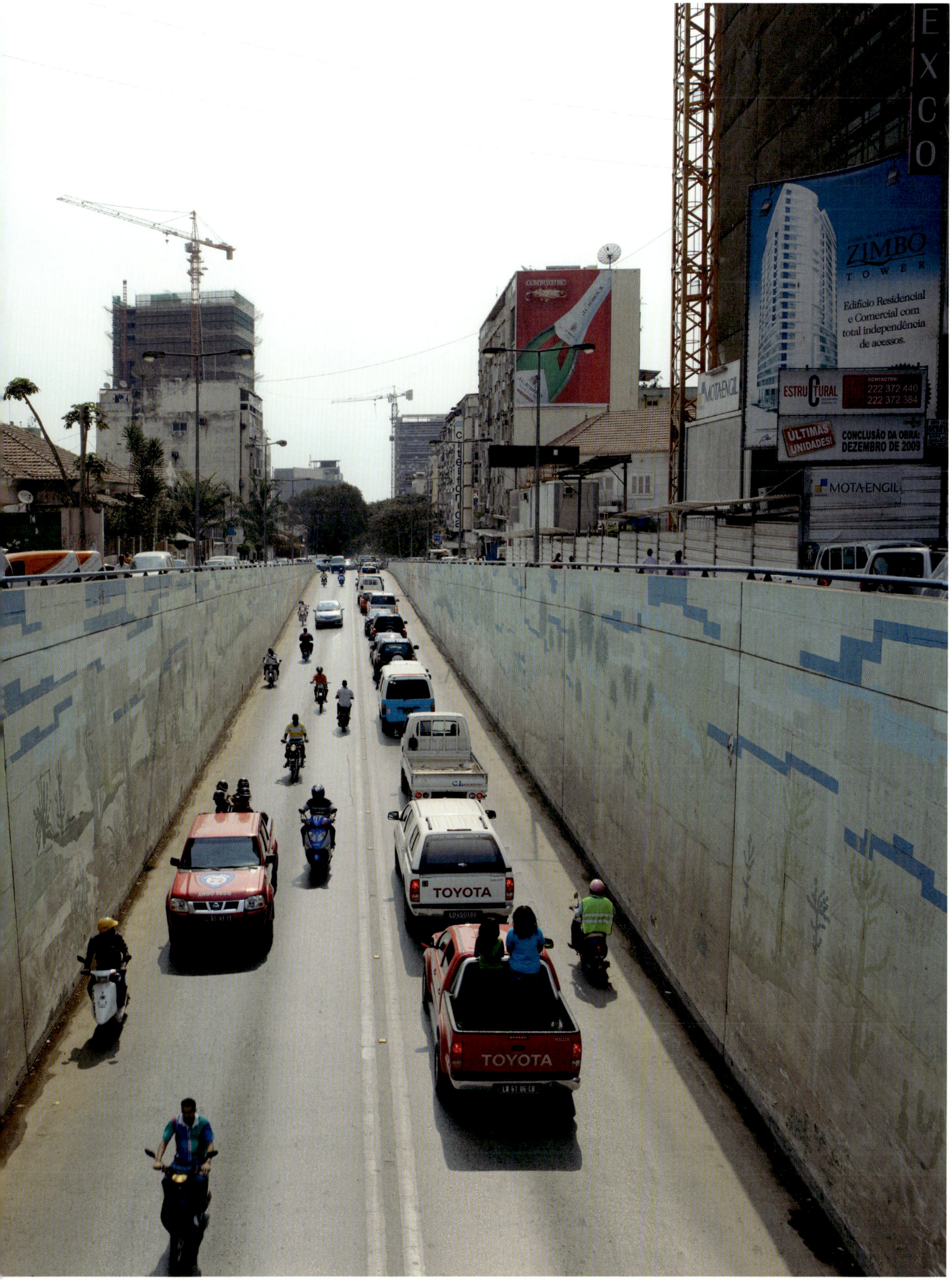
EXCO
ZIMBO
TOWER
Edifício Residencial e Comercial com total independência de acessos.
ESTRUCTURAL
CONTACTOS:
222 372 440
222 372 384
ÚLTIMAS UNIDADES!
CONCLUSÃO DA OBRA: DEZEMBRO DE 2009
MOTA-ENGIL
MOTA-ENGIL
TOYOTA
TOYOTA

ASTANA STREET
CAPE TOWN

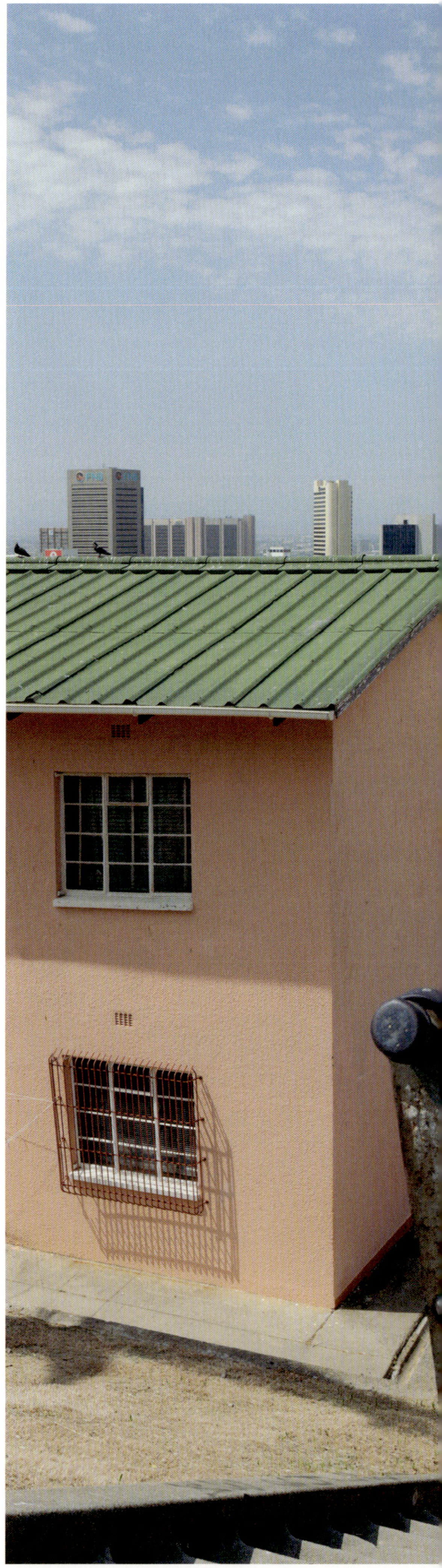

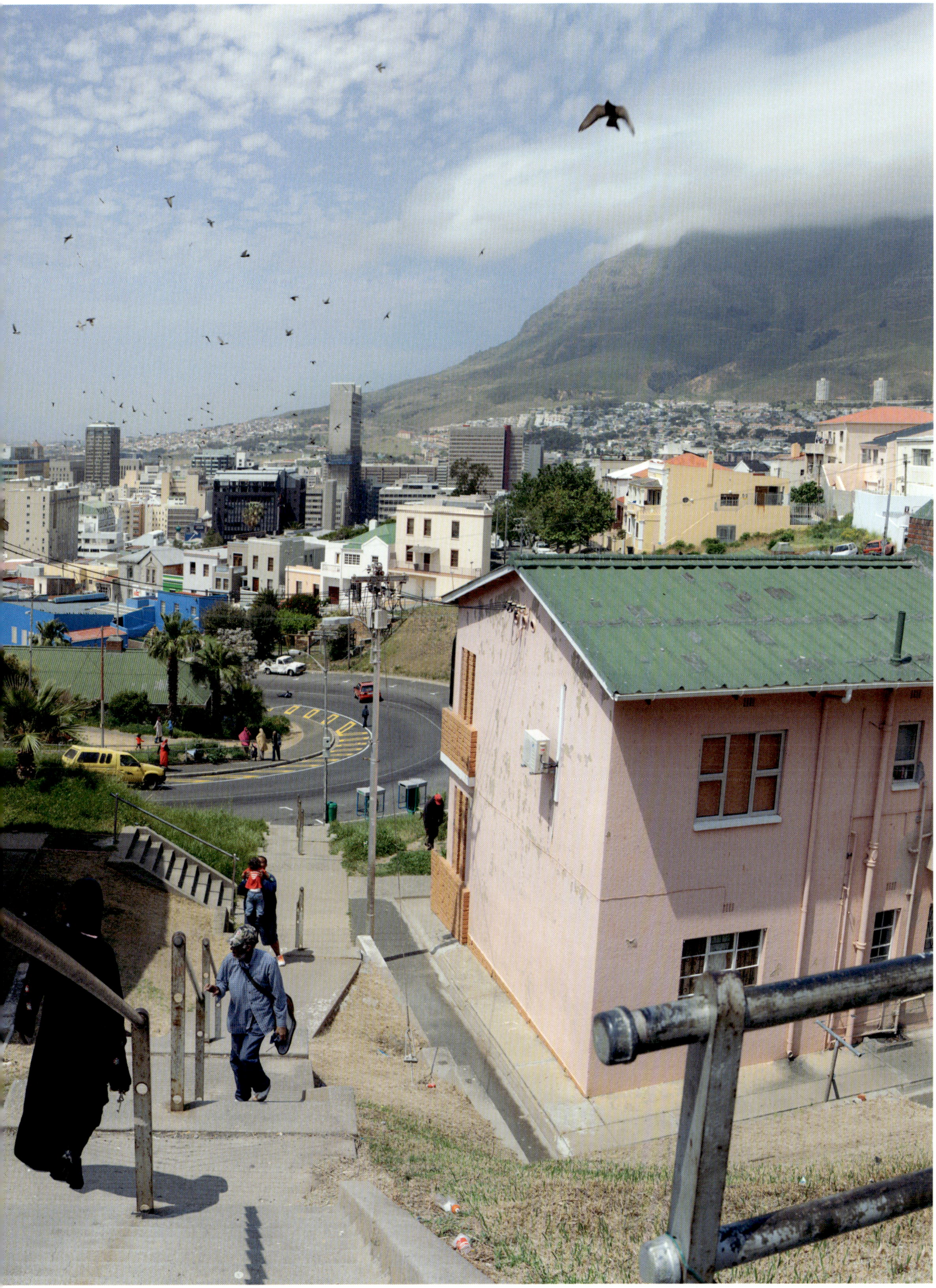

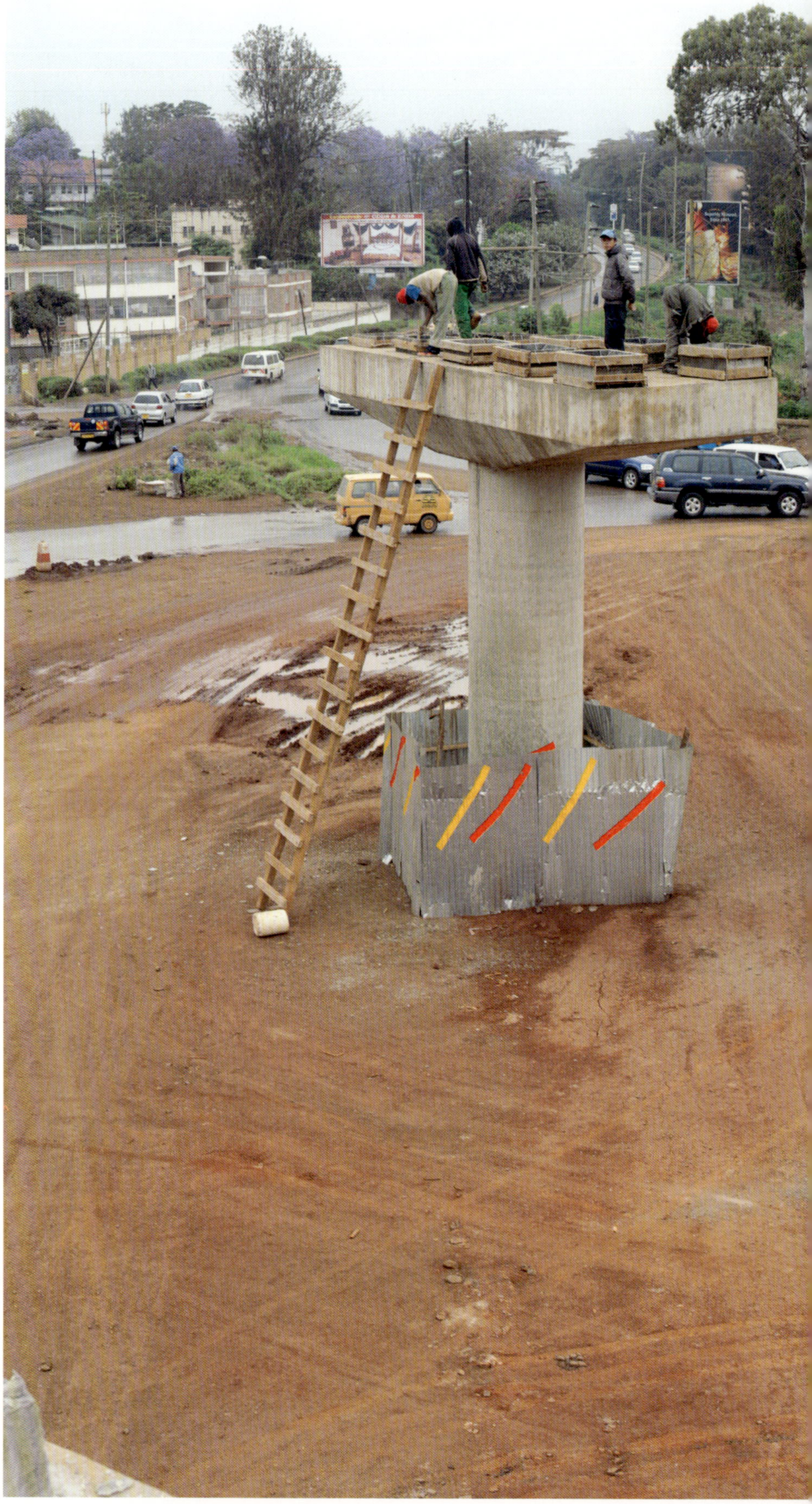

FOREST ROAD & LIMURU ROAD
NAIROBI

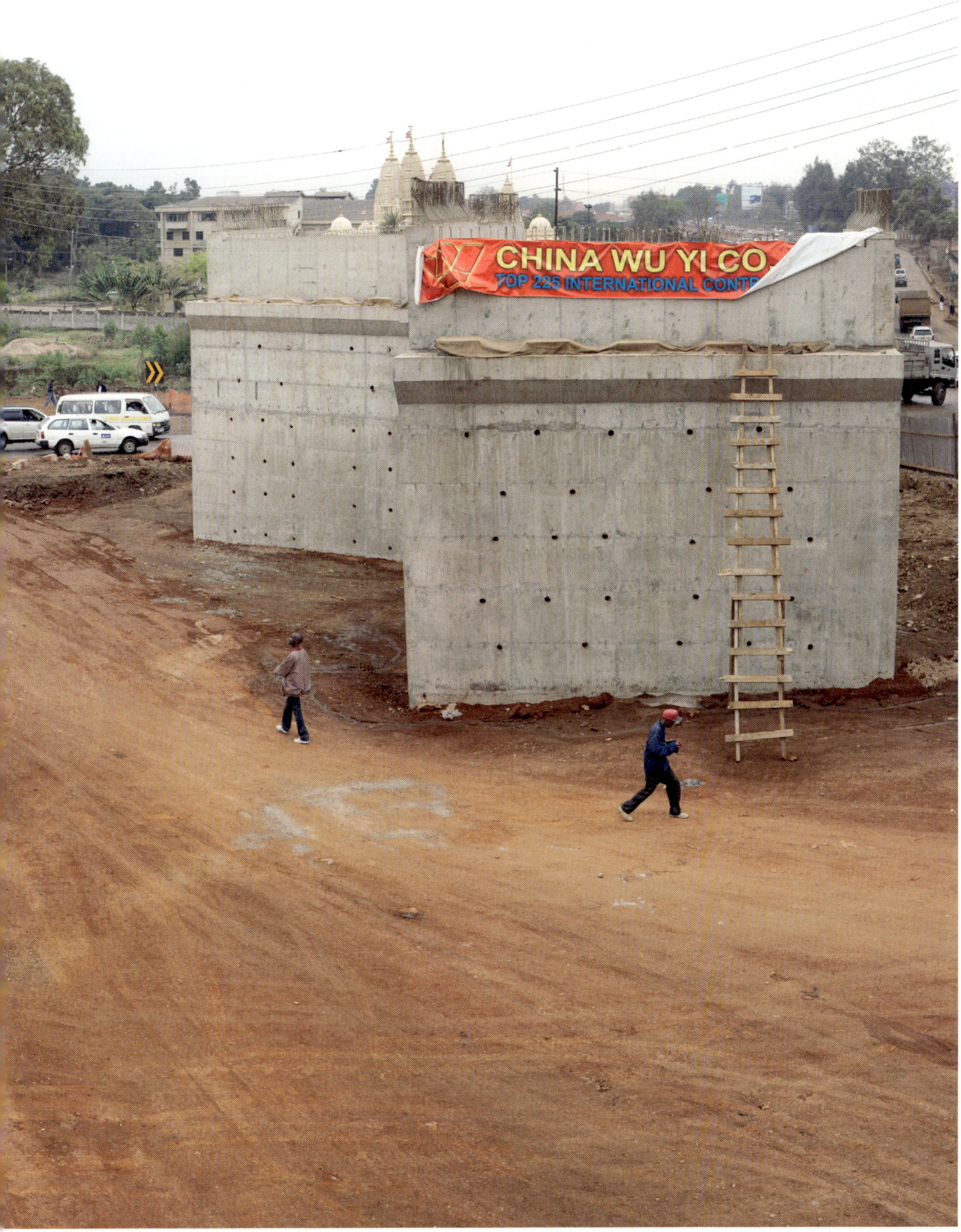
CHINA WU YI CO.
TOP 225 INTERNATIONAL CONTR

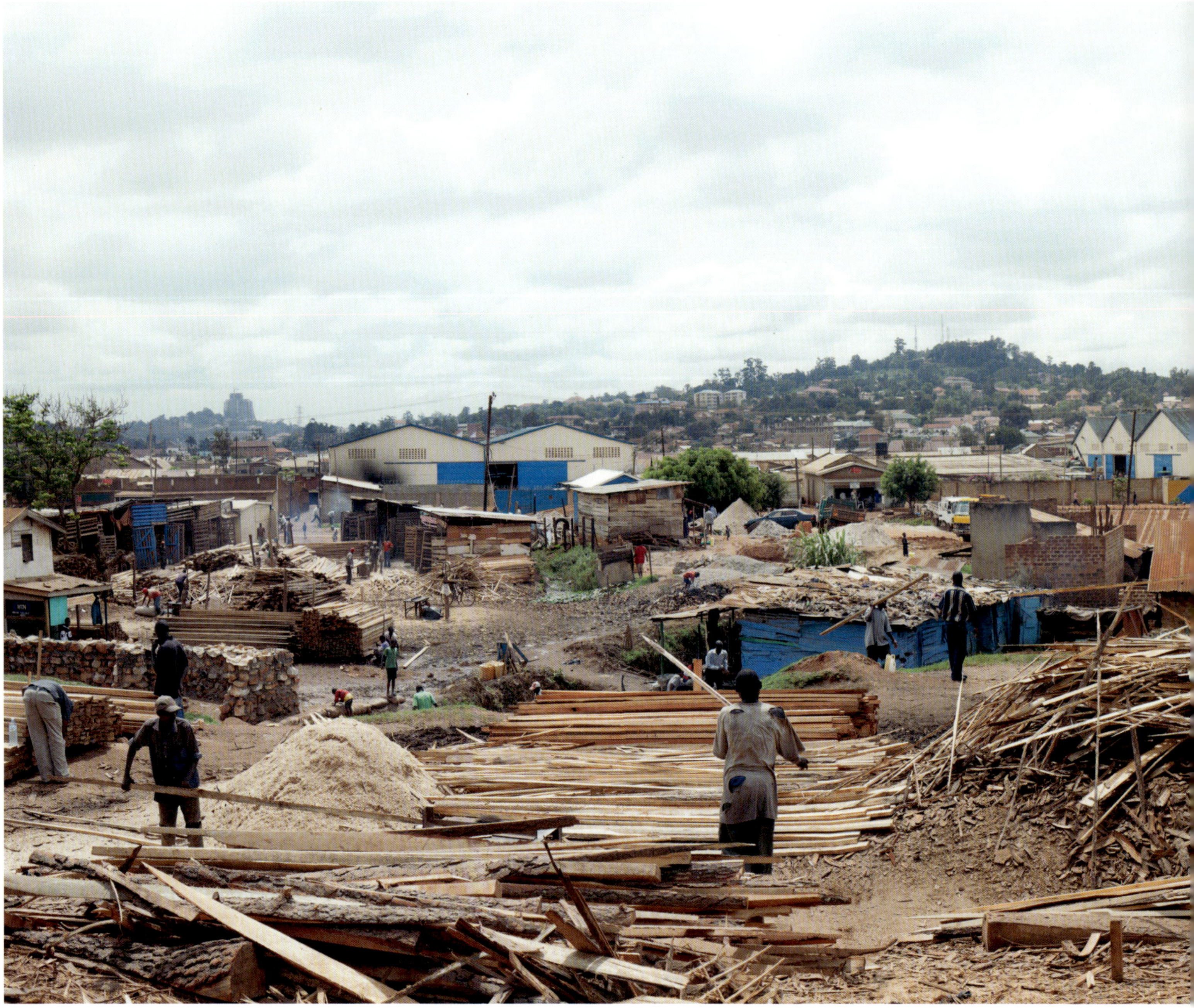

BWAISE (LUMBERYARD)
KAMPALA

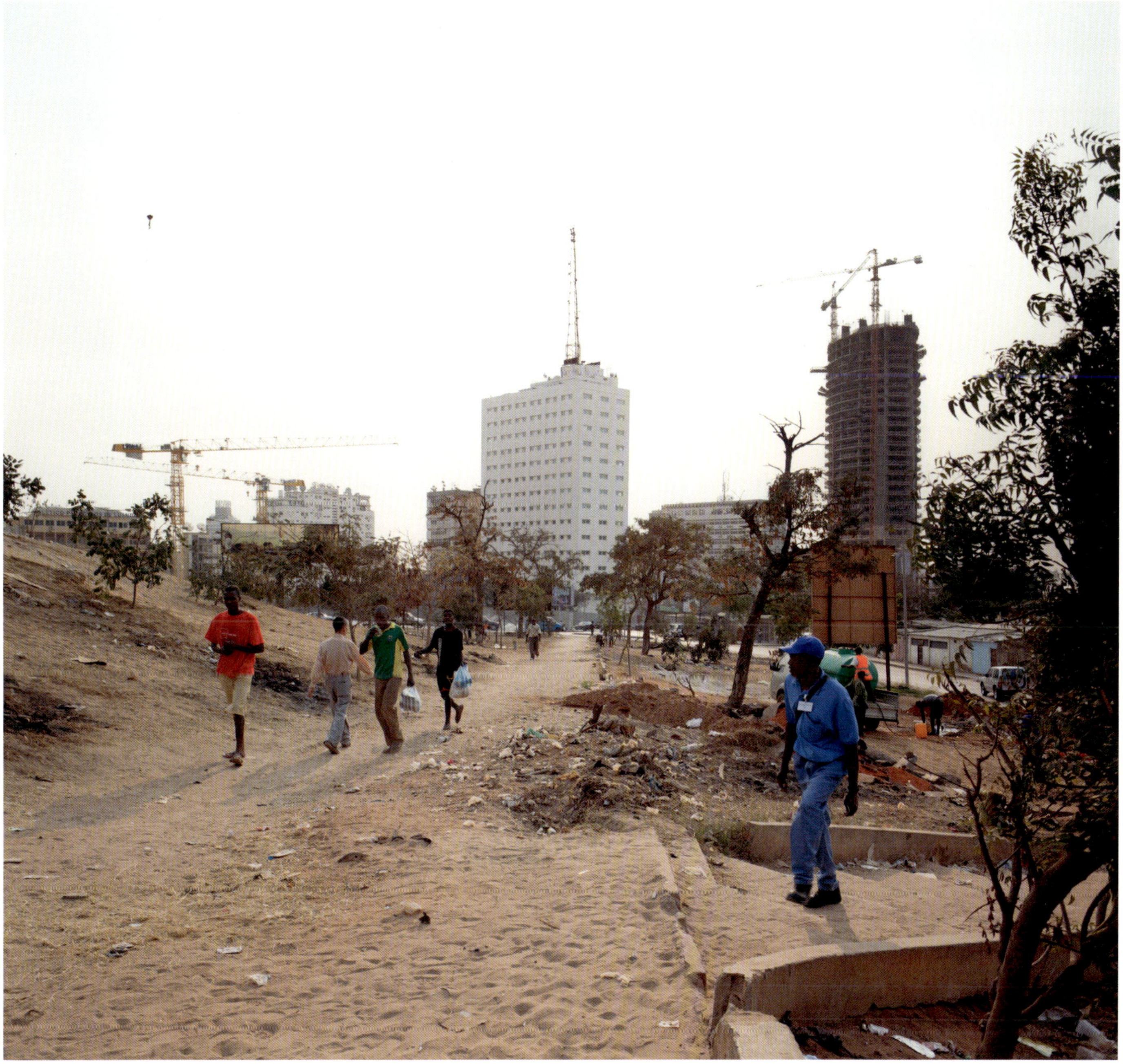

AVENIDA NEHRU
LUANDA

DALAJANI, MANZESE
DAR ES SALAAM

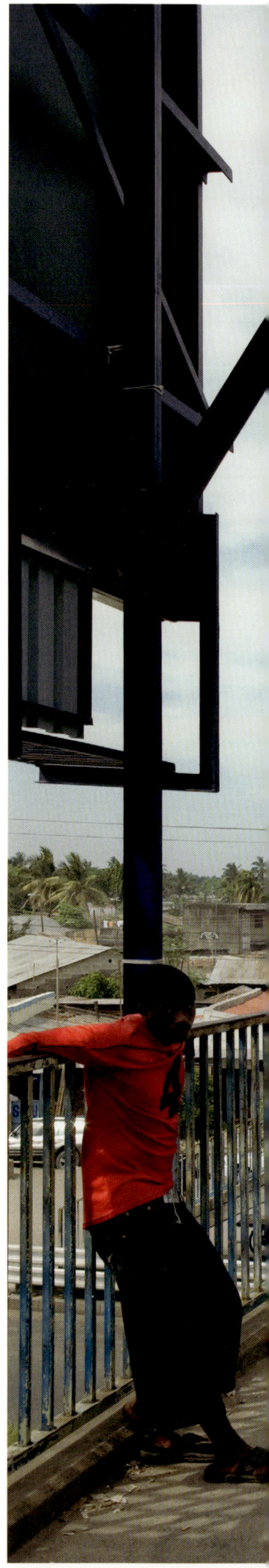

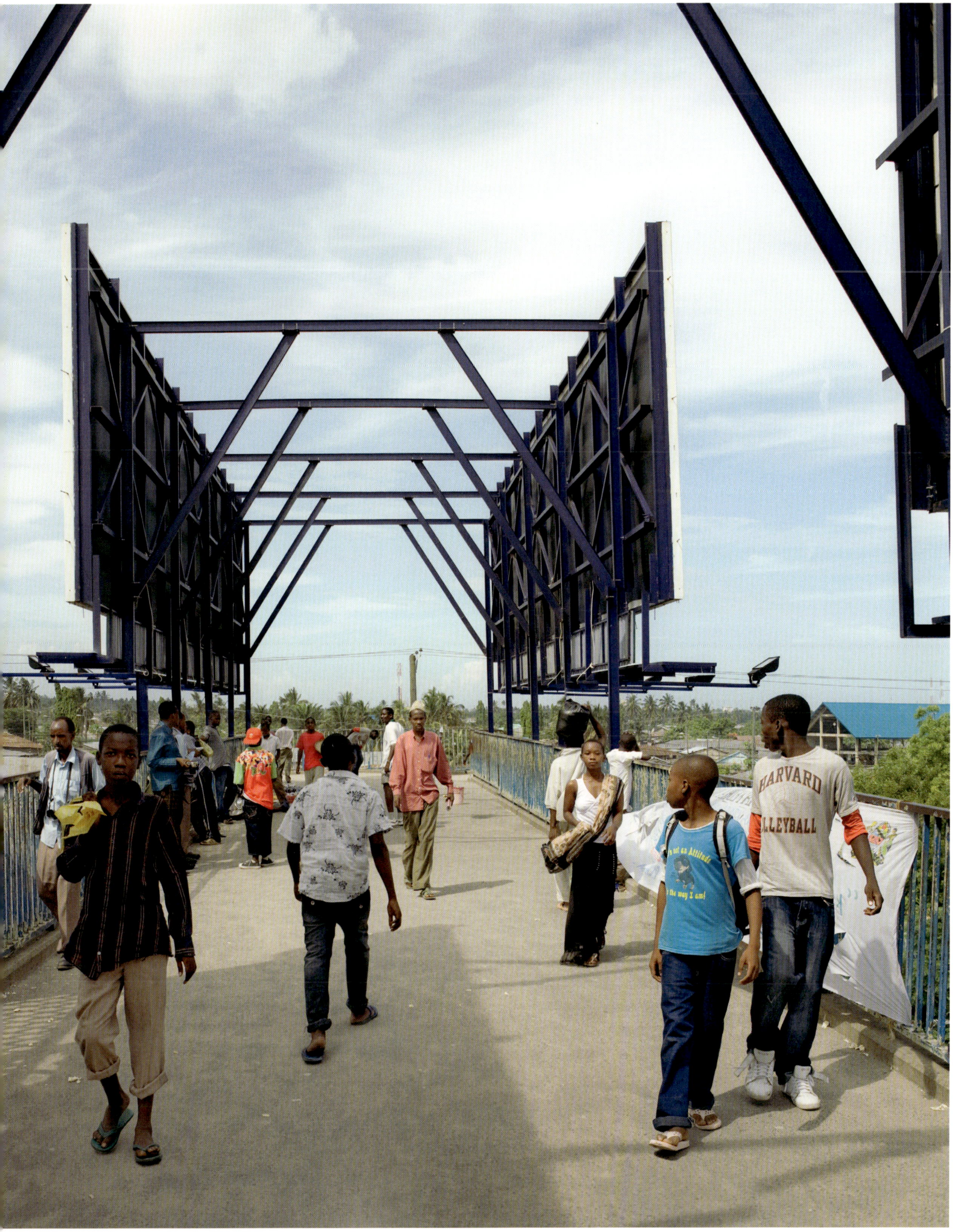
HARVARD
LLEYBALL

KARIAKOO MARKET
DAR ES SALAAM

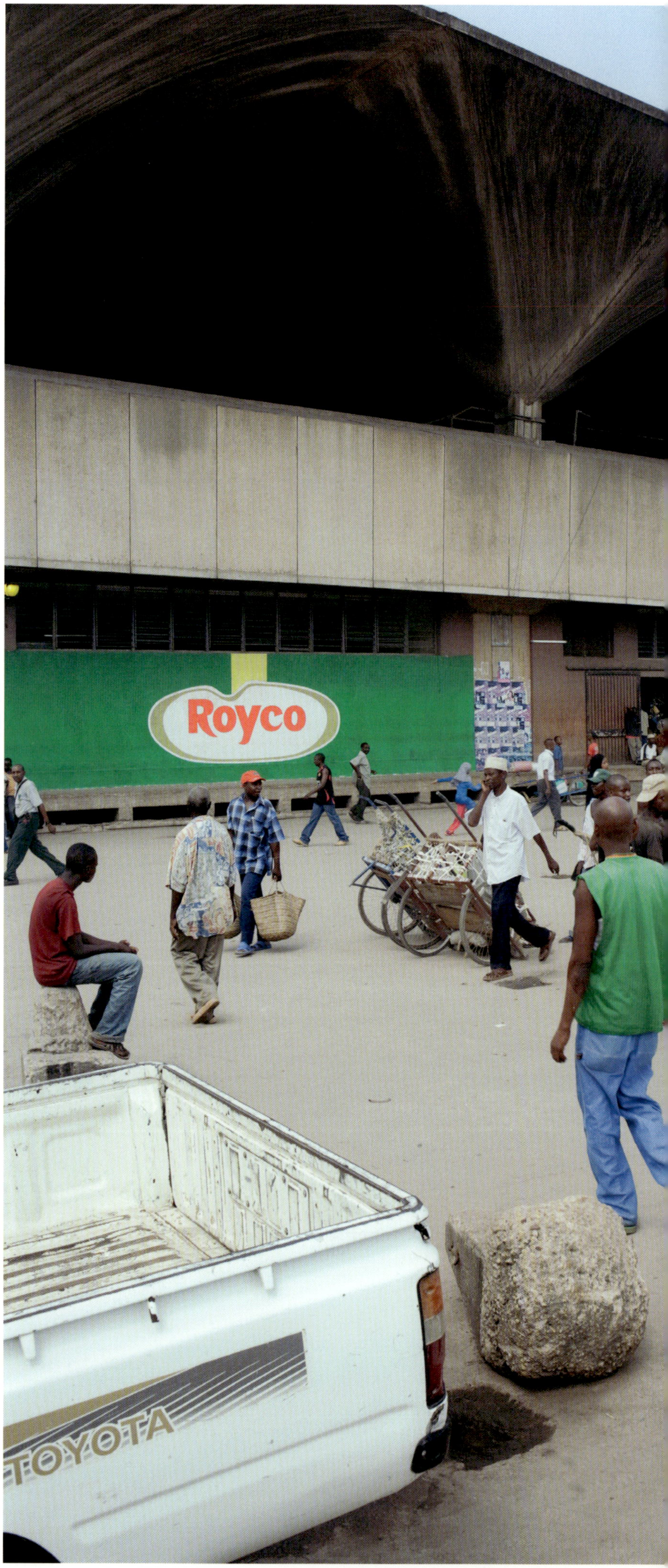

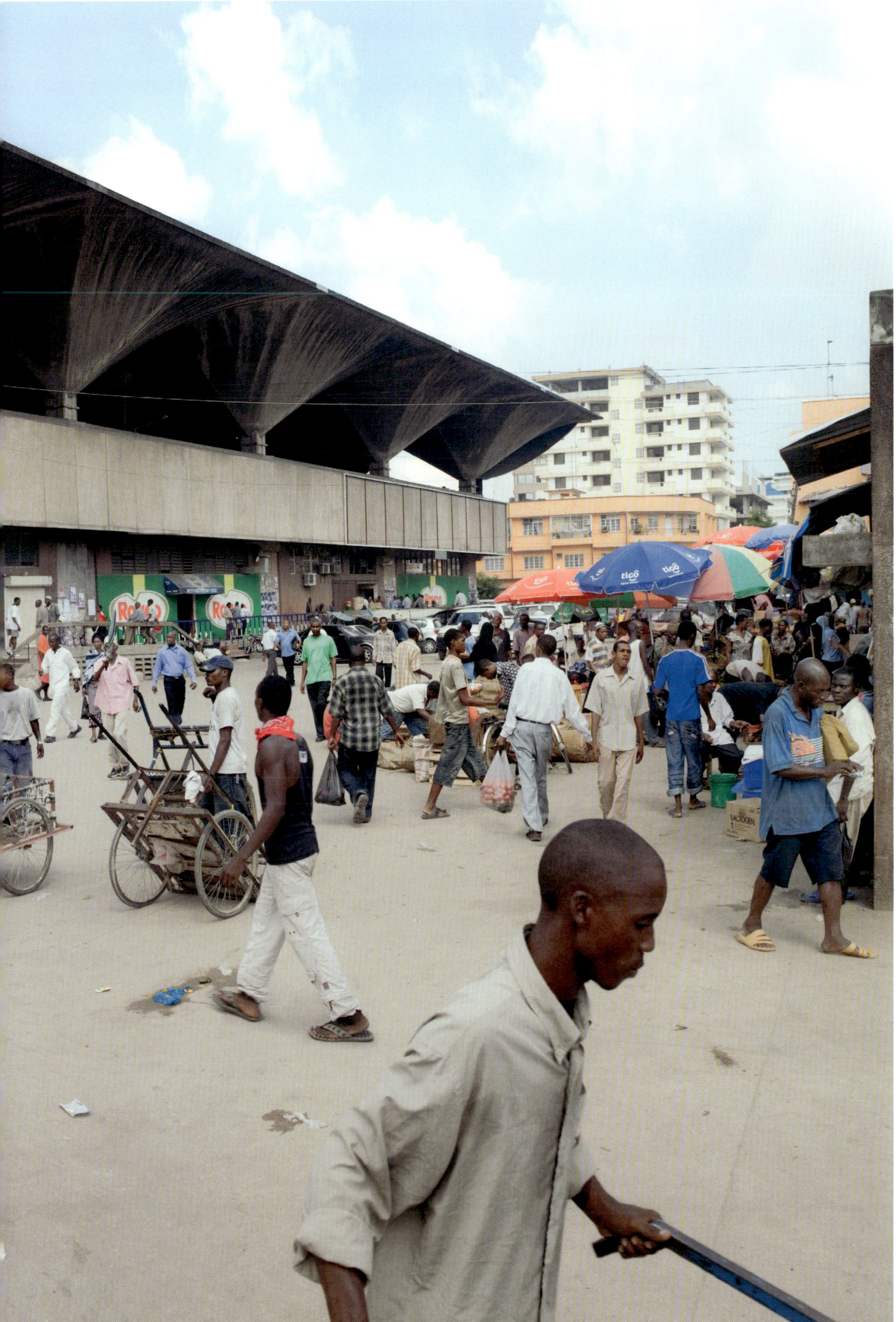
Royco
Royco
tigo
tigo

ONIRU ESTATE, VICTORIA ISLAND
LAGOS

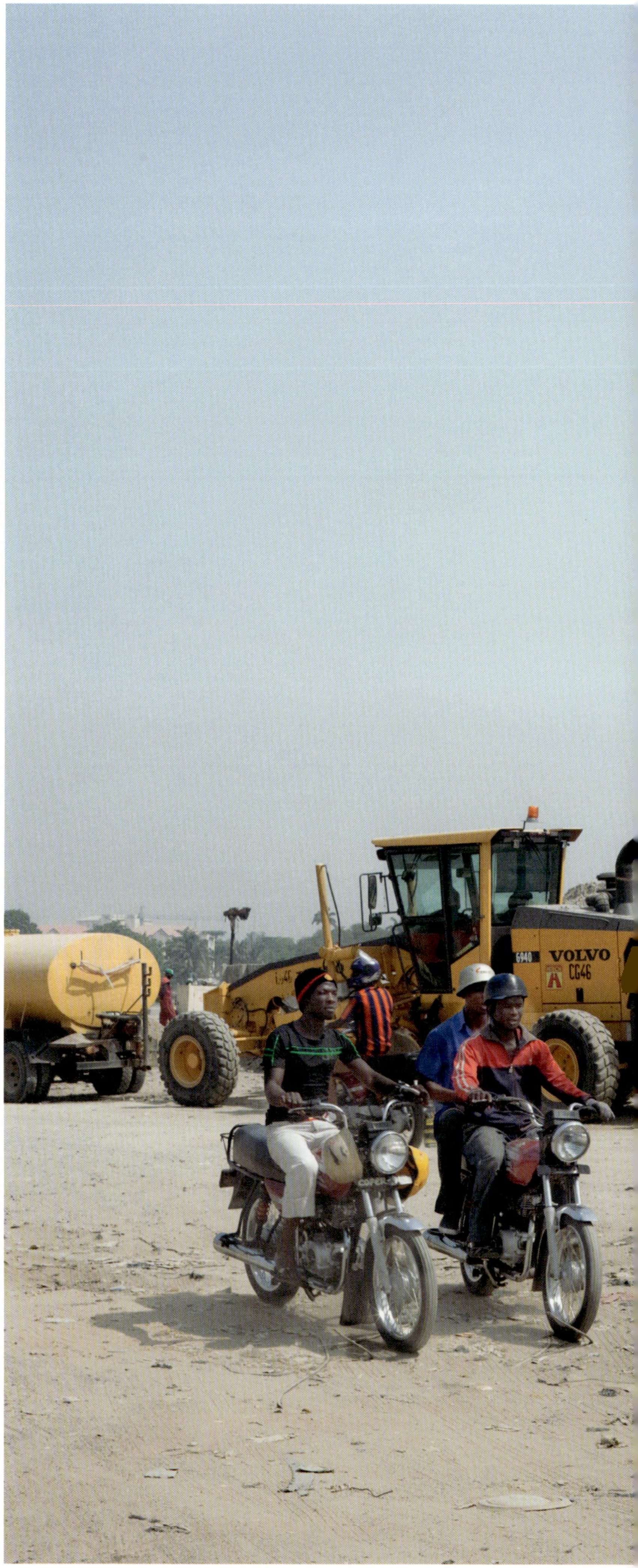

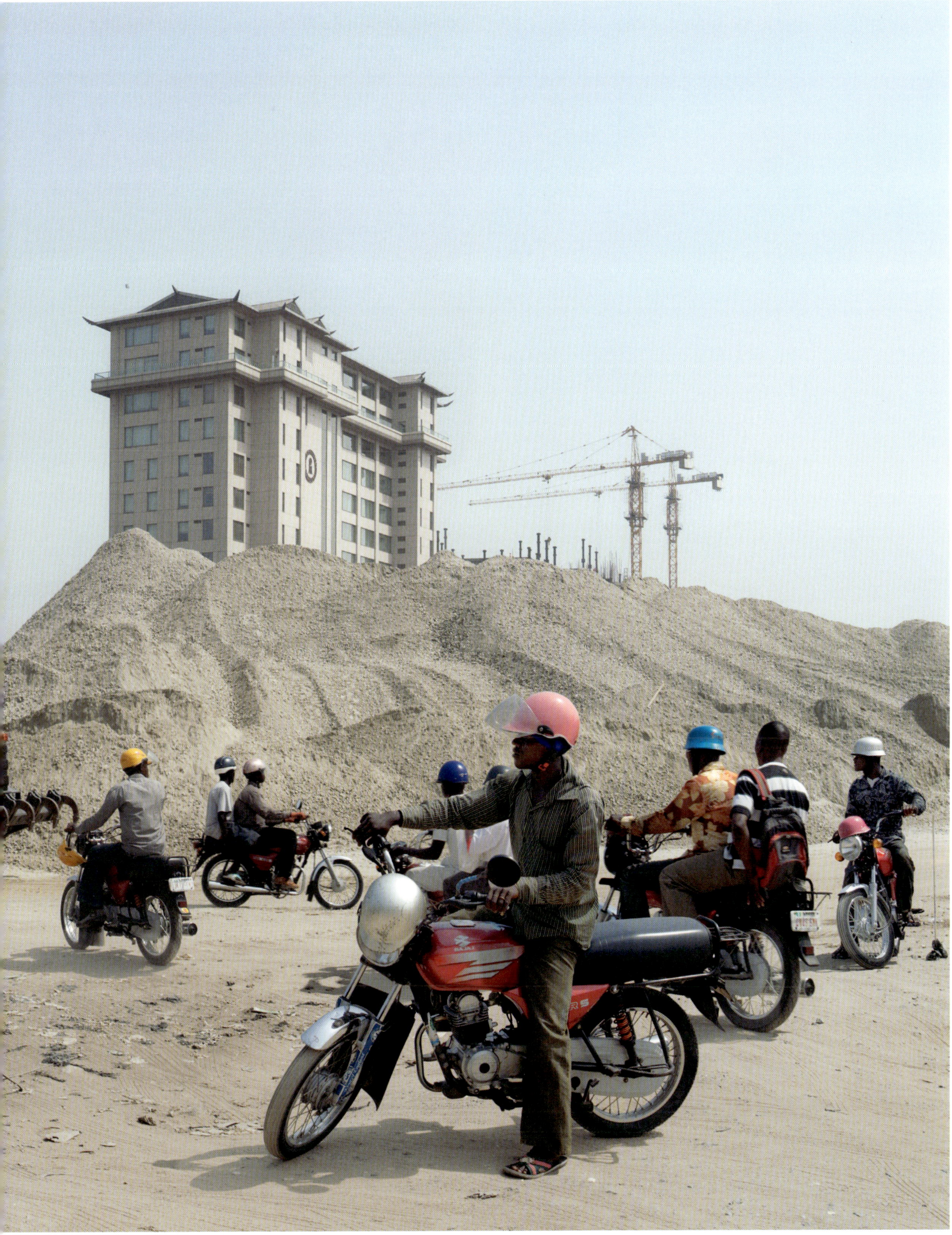

HADFIELD ROAD, HILLBROW
JOHANNESBURG

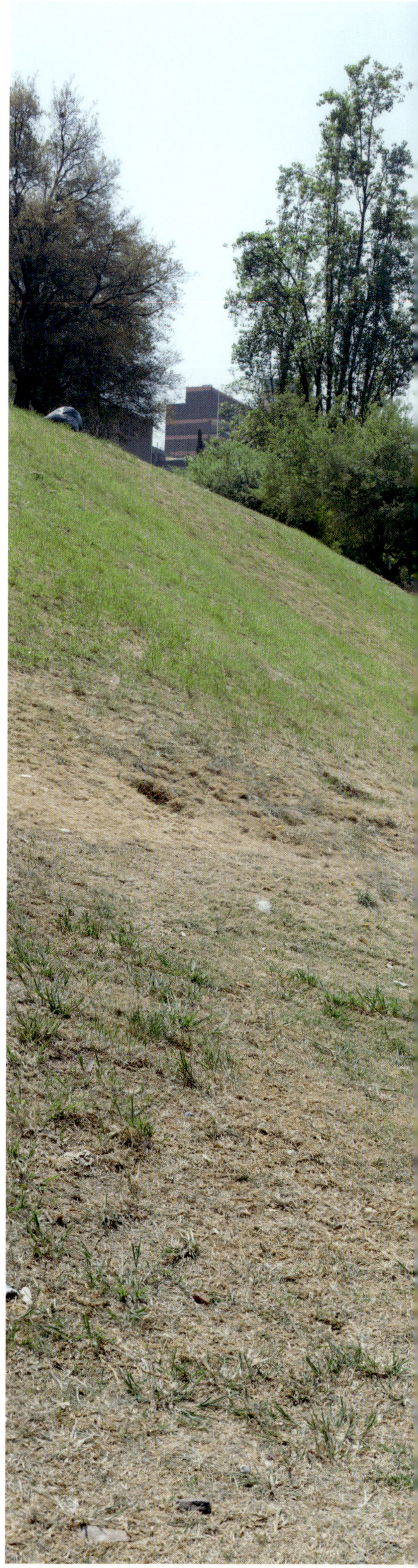

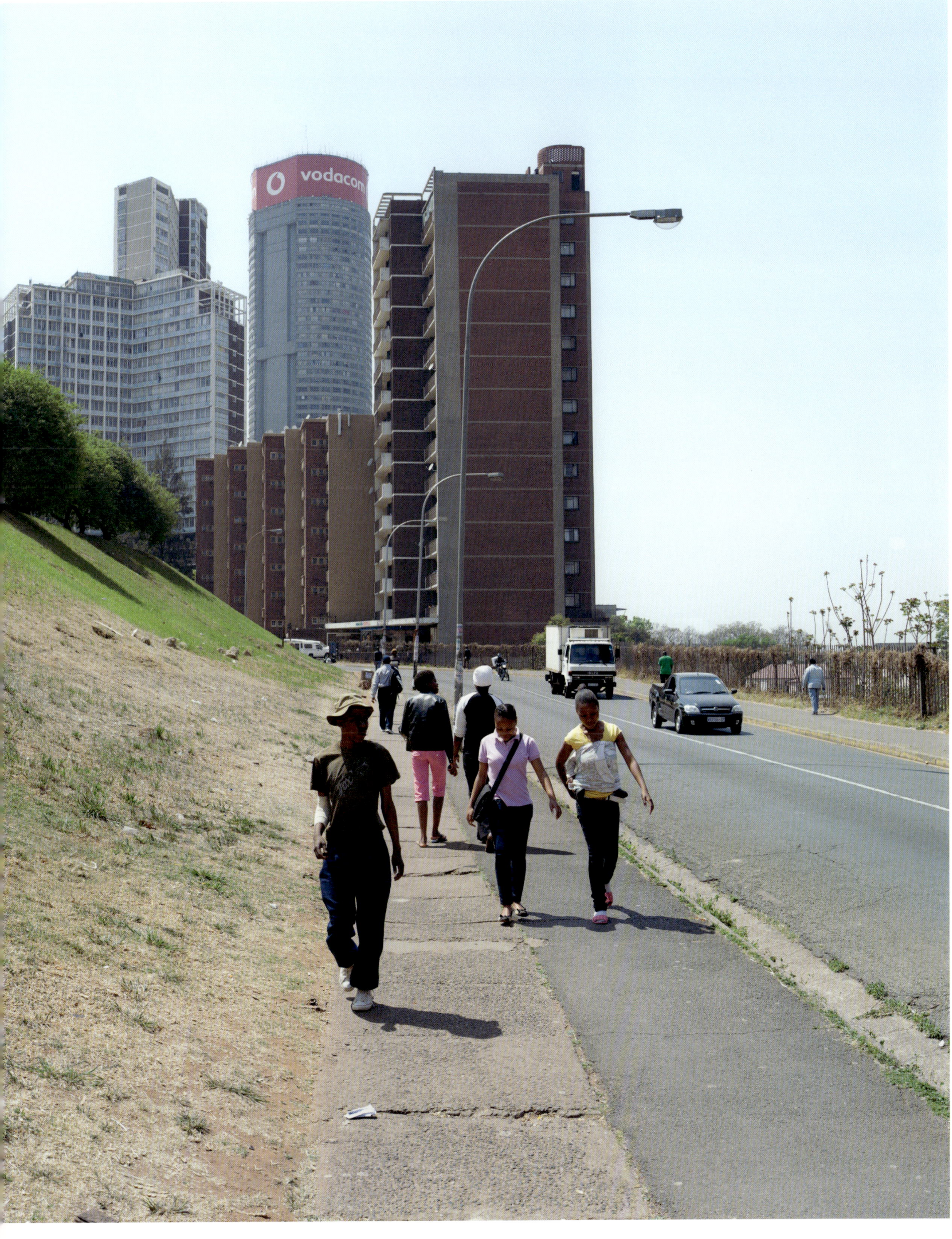
vodacom

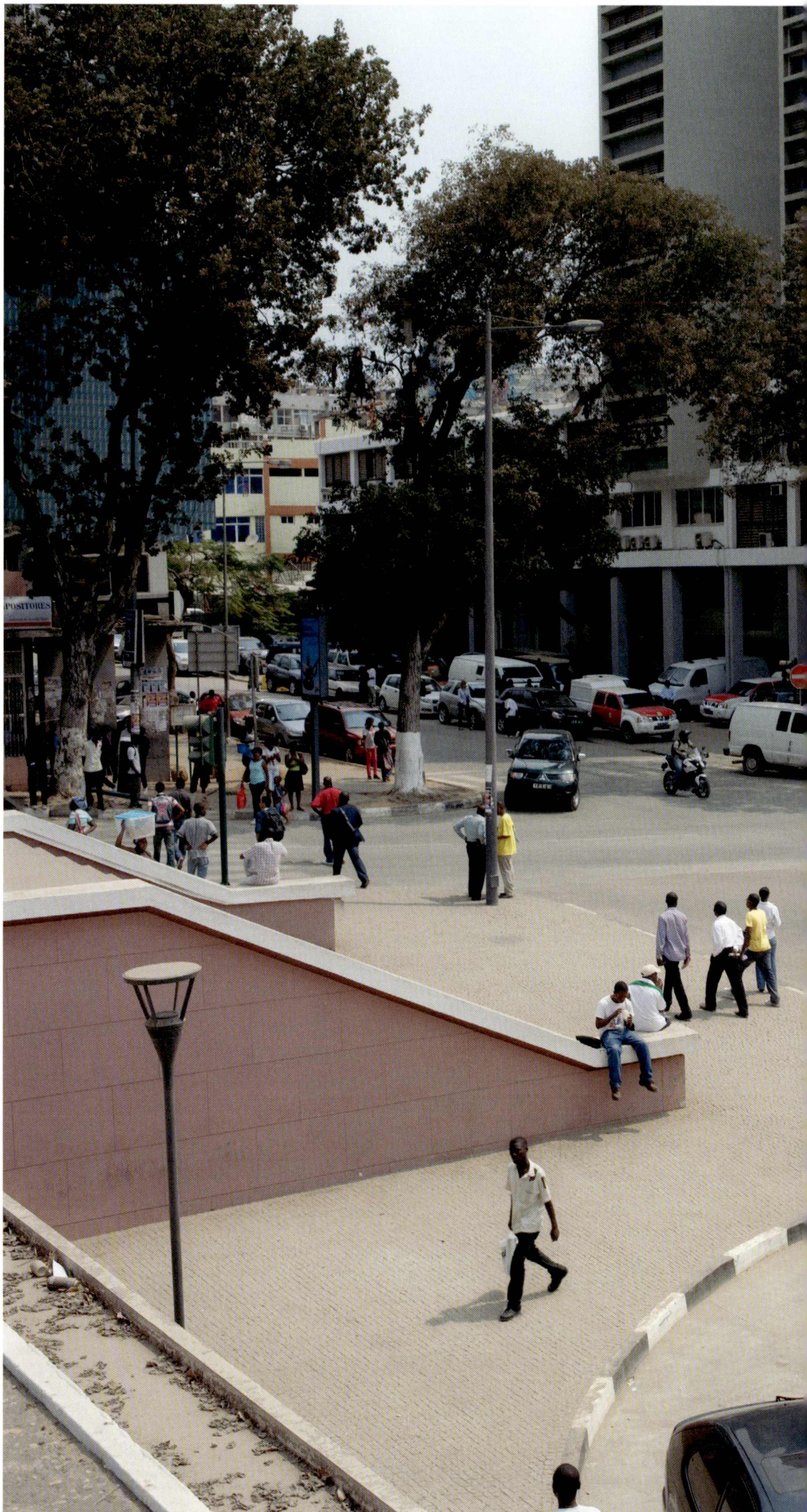
POSITORES

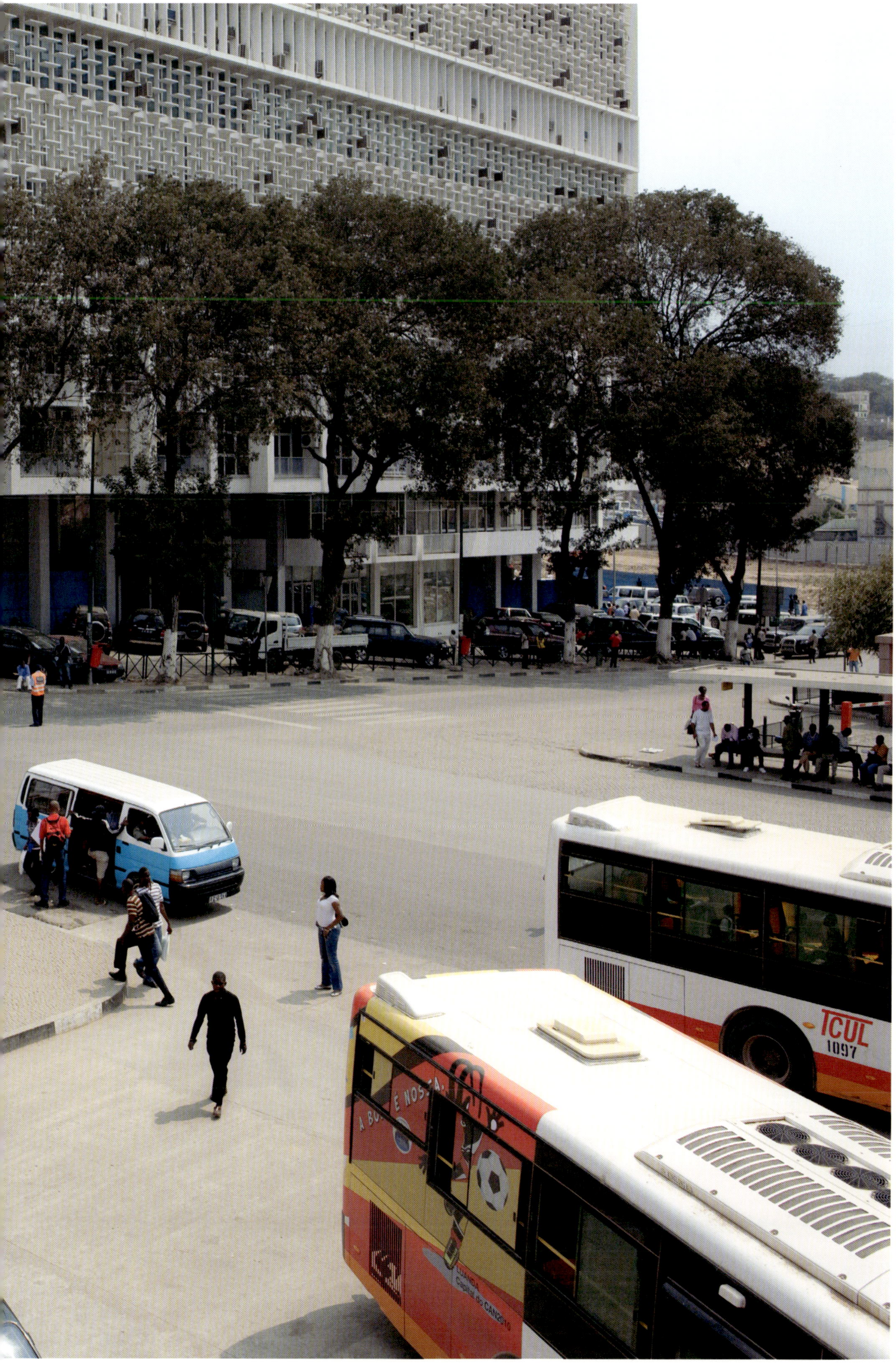
TCUL
1097
LUANDA
Capital do CAN2010

RUE 14, HAY AL BARAKA
CASABLANCA

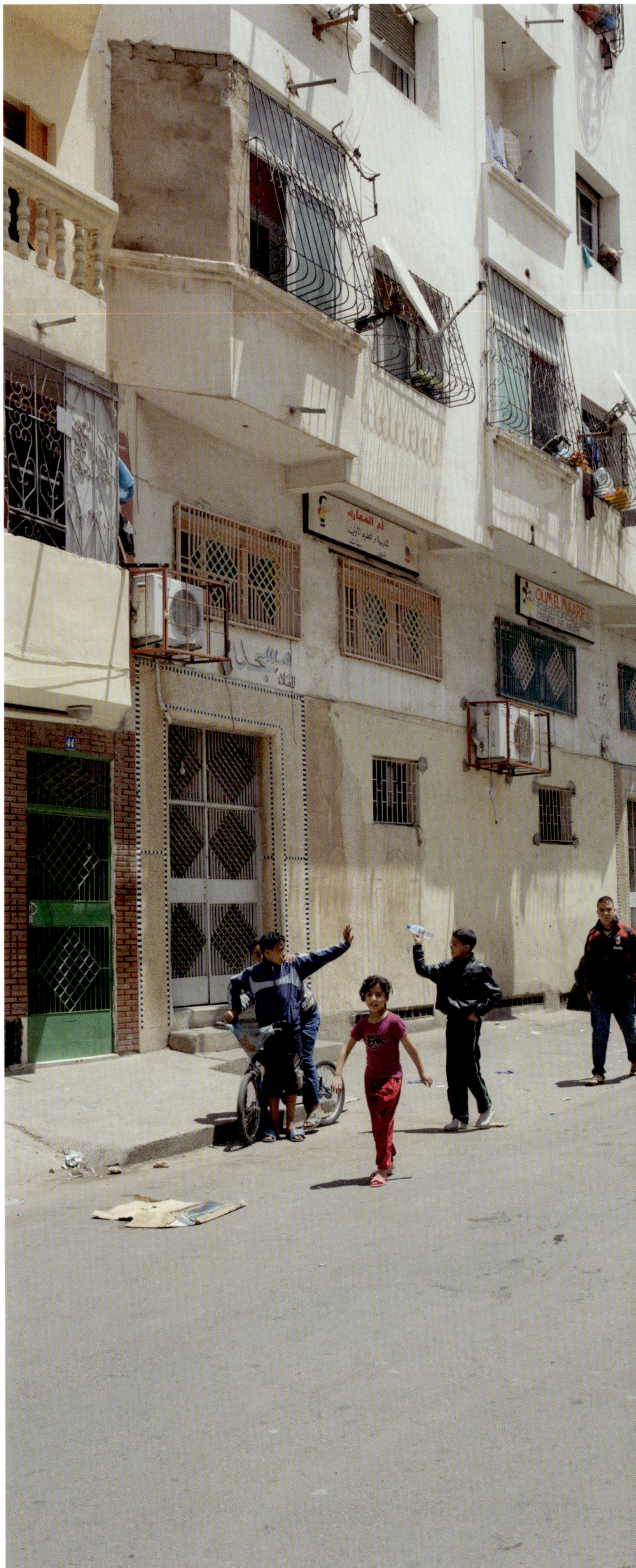

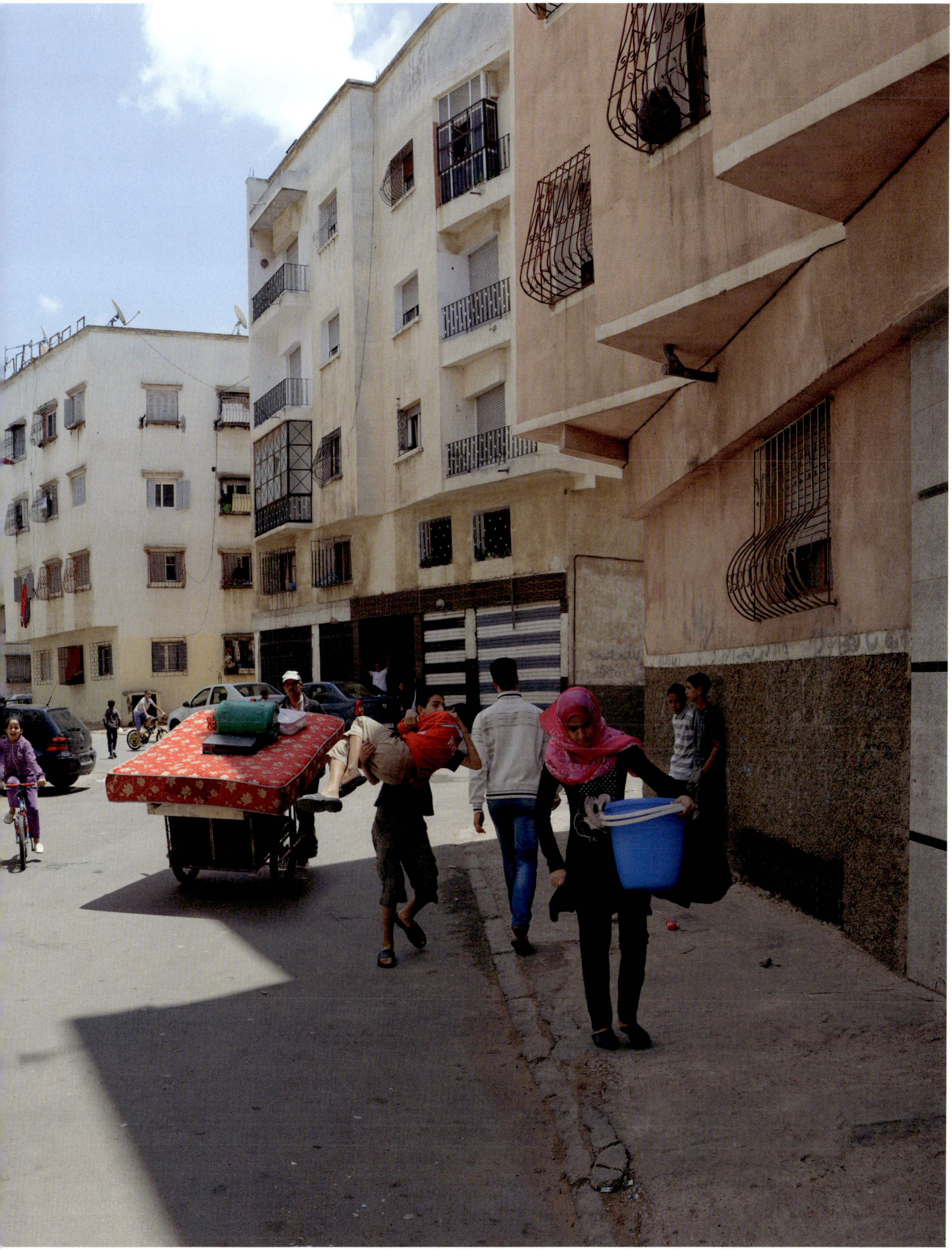

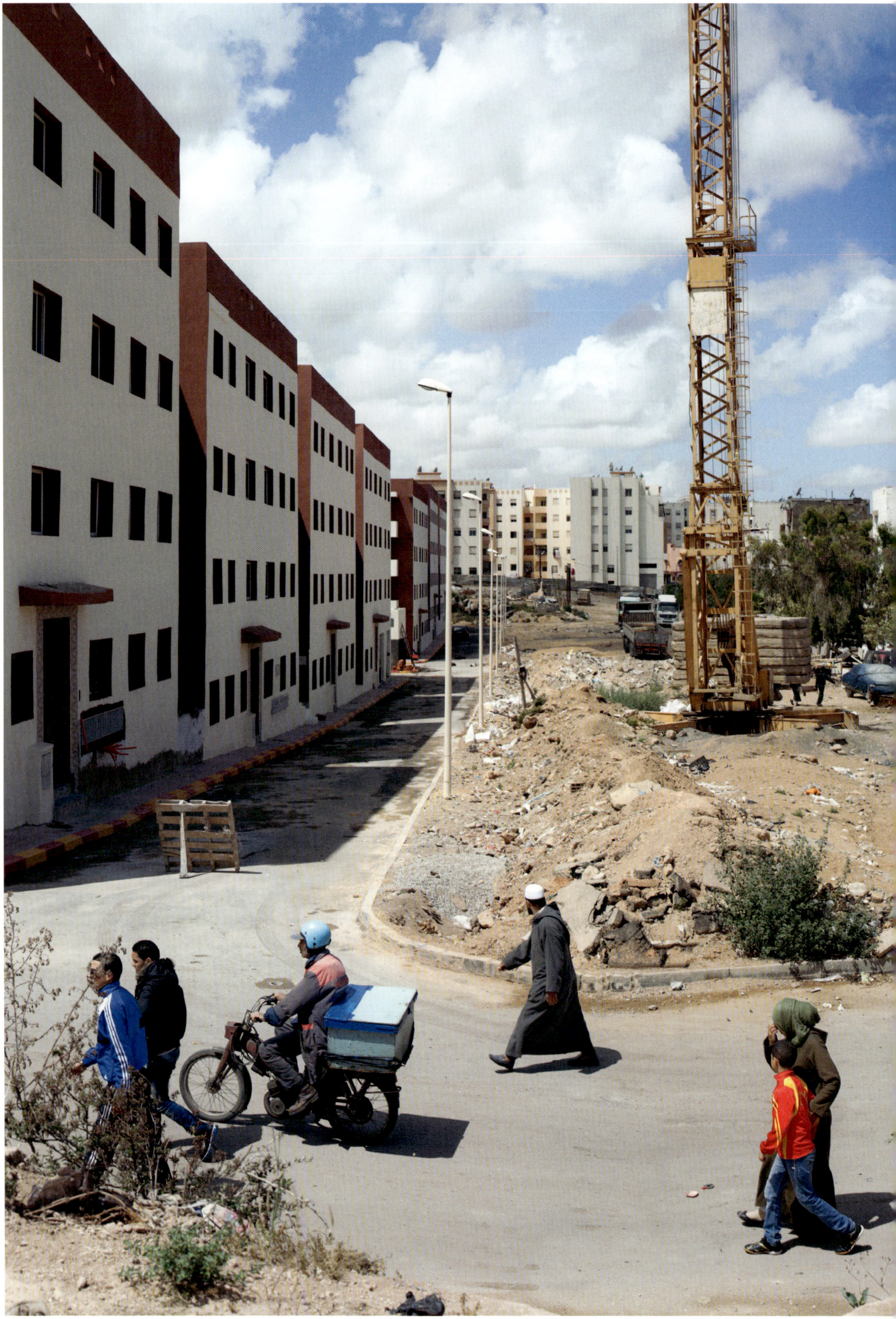

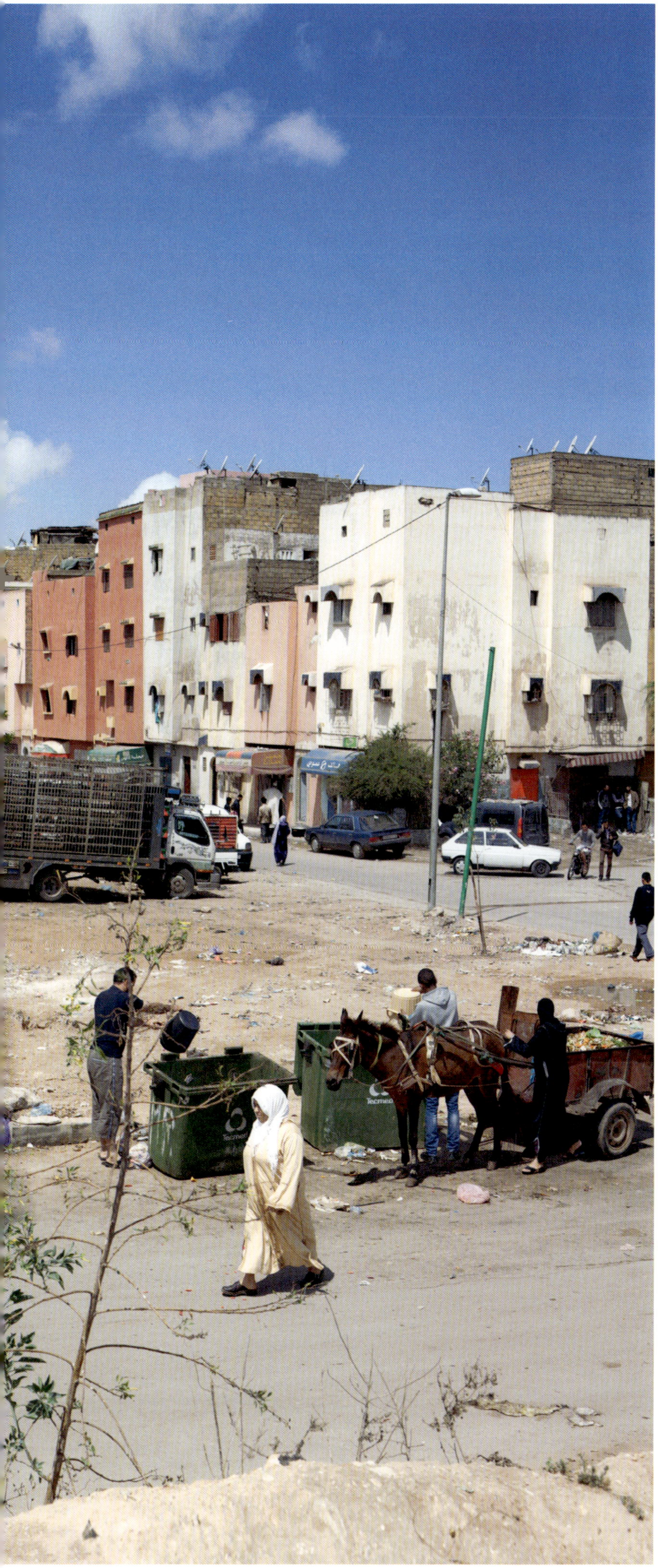

HAY AL HOUDA
CASABLANCA

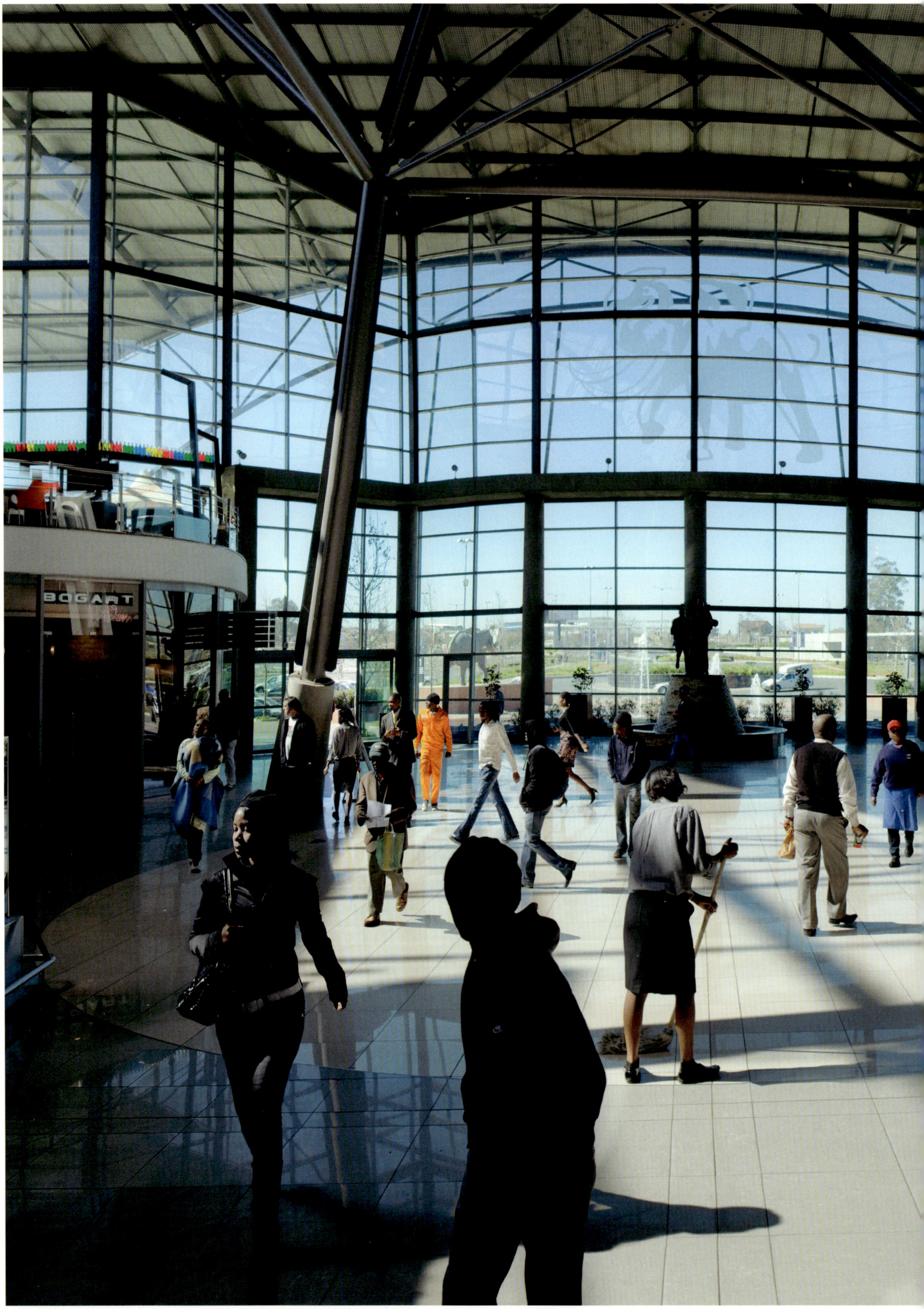
BOGART

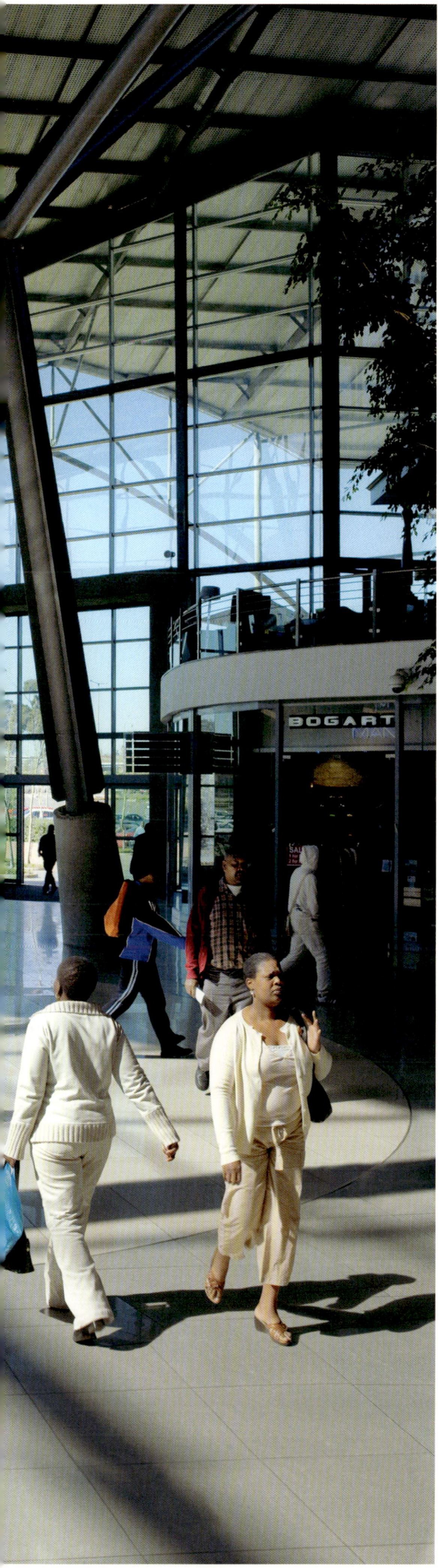

MAPUNYA MALL, SOWETO
JOHANNESBURG

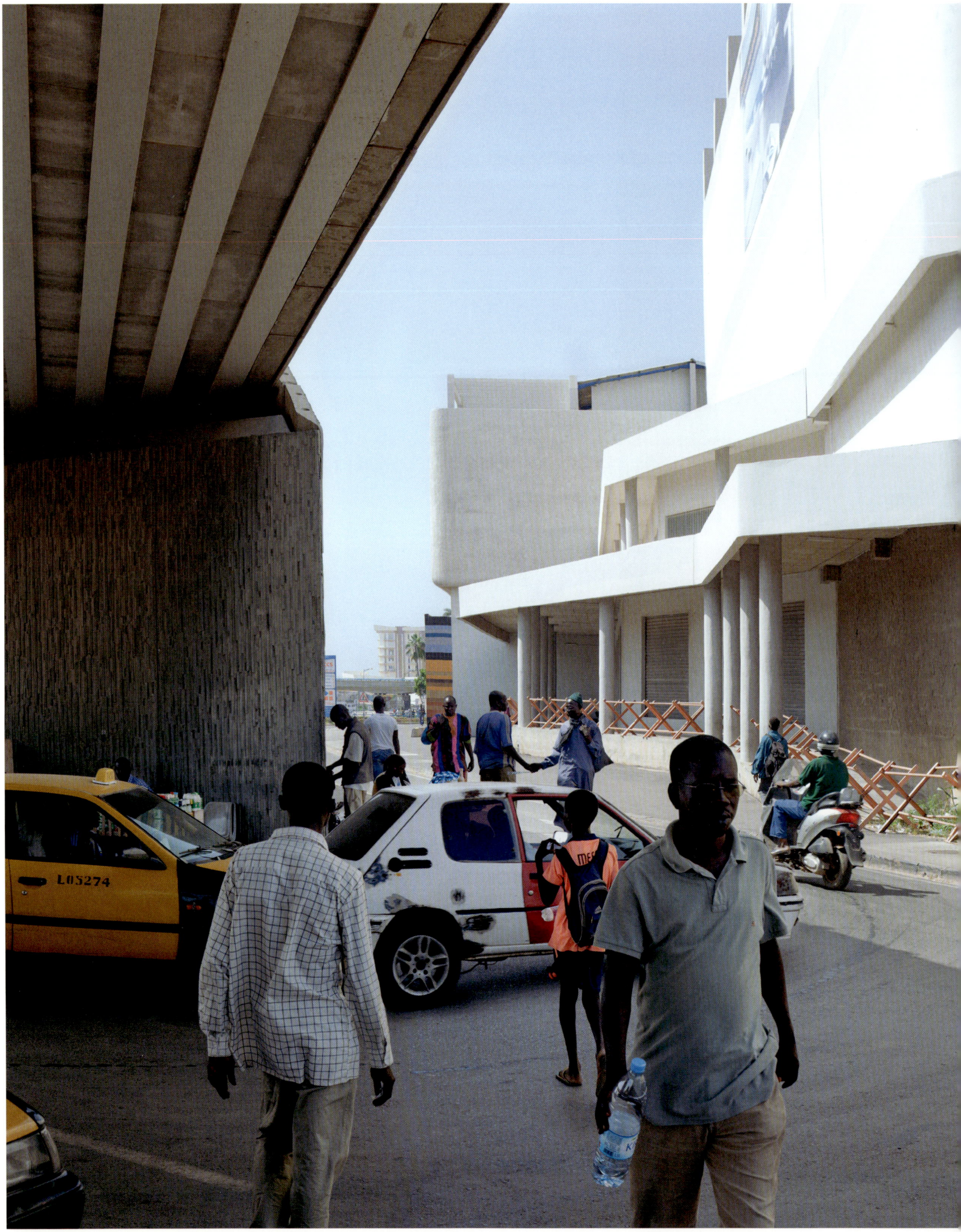
L05274

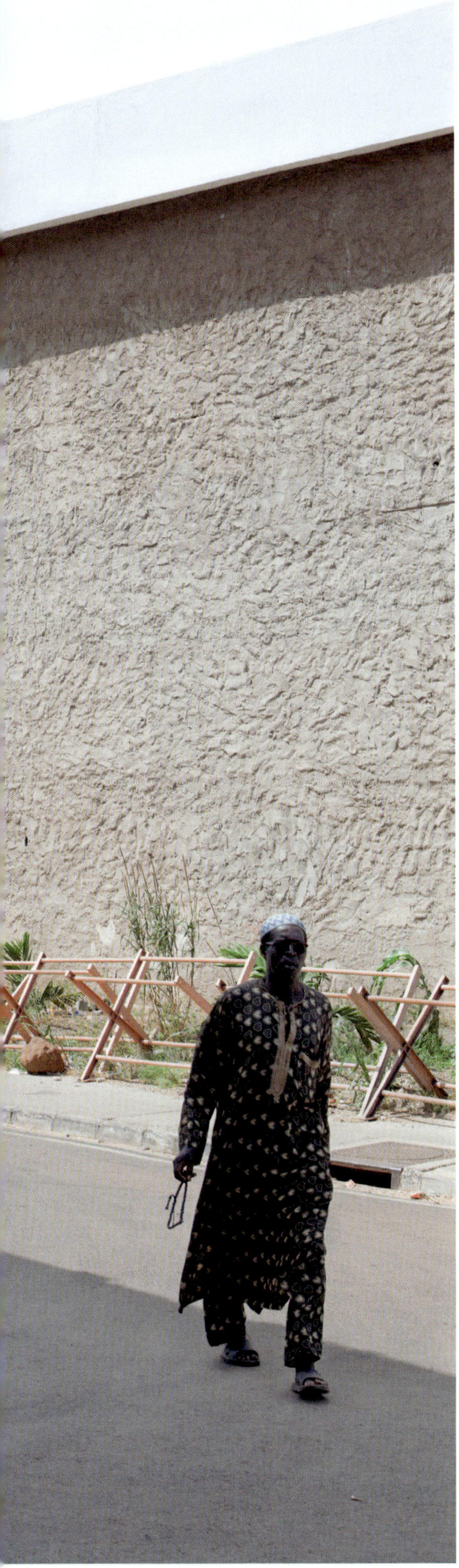

PLACE CYRNOS (C.C.S.)
DAKAR

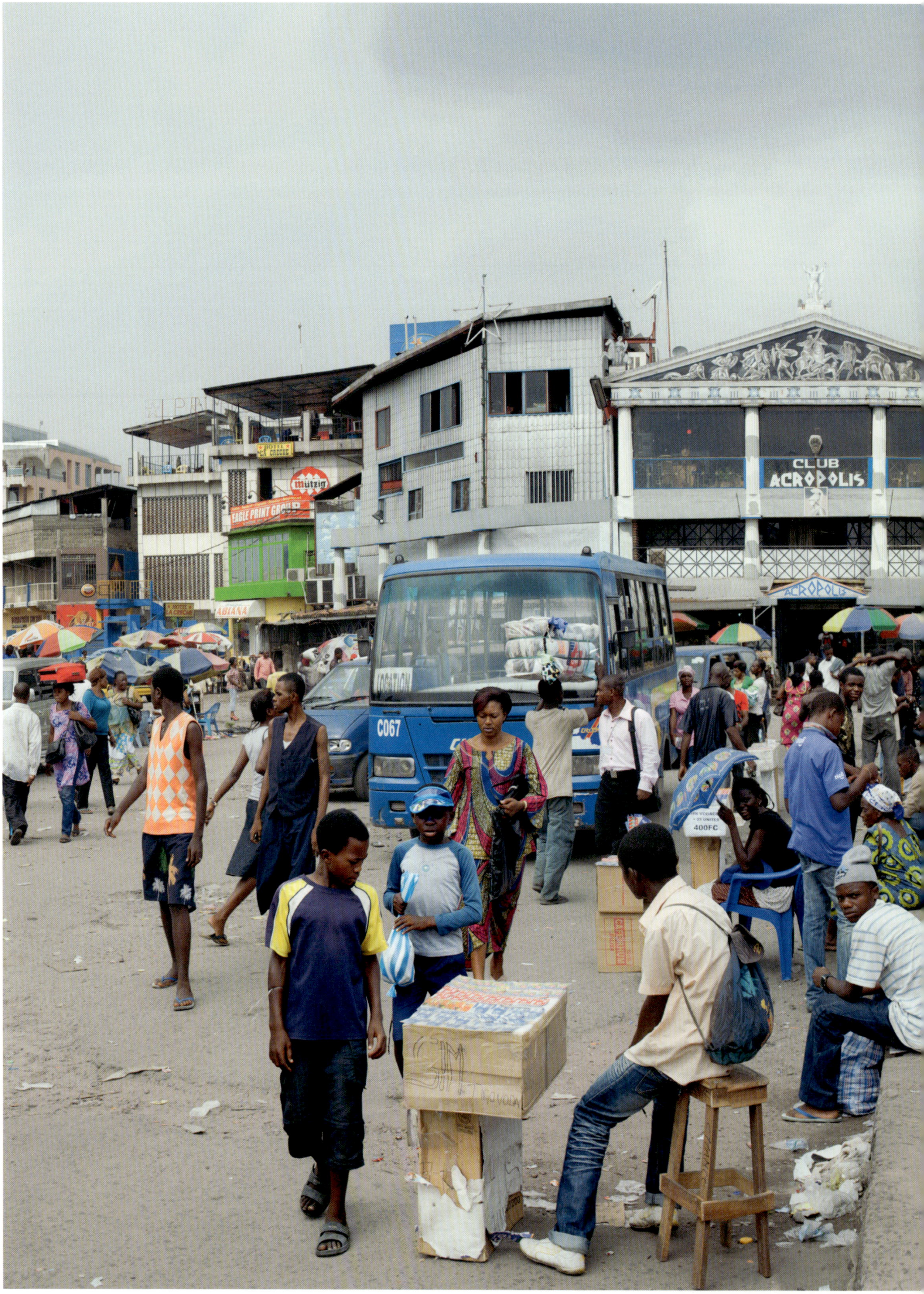
CLUB
ACROPOLIS
ACROPOLIS
EAGLE PRINT GROUP
mützig
C067
LOCATION
400FC

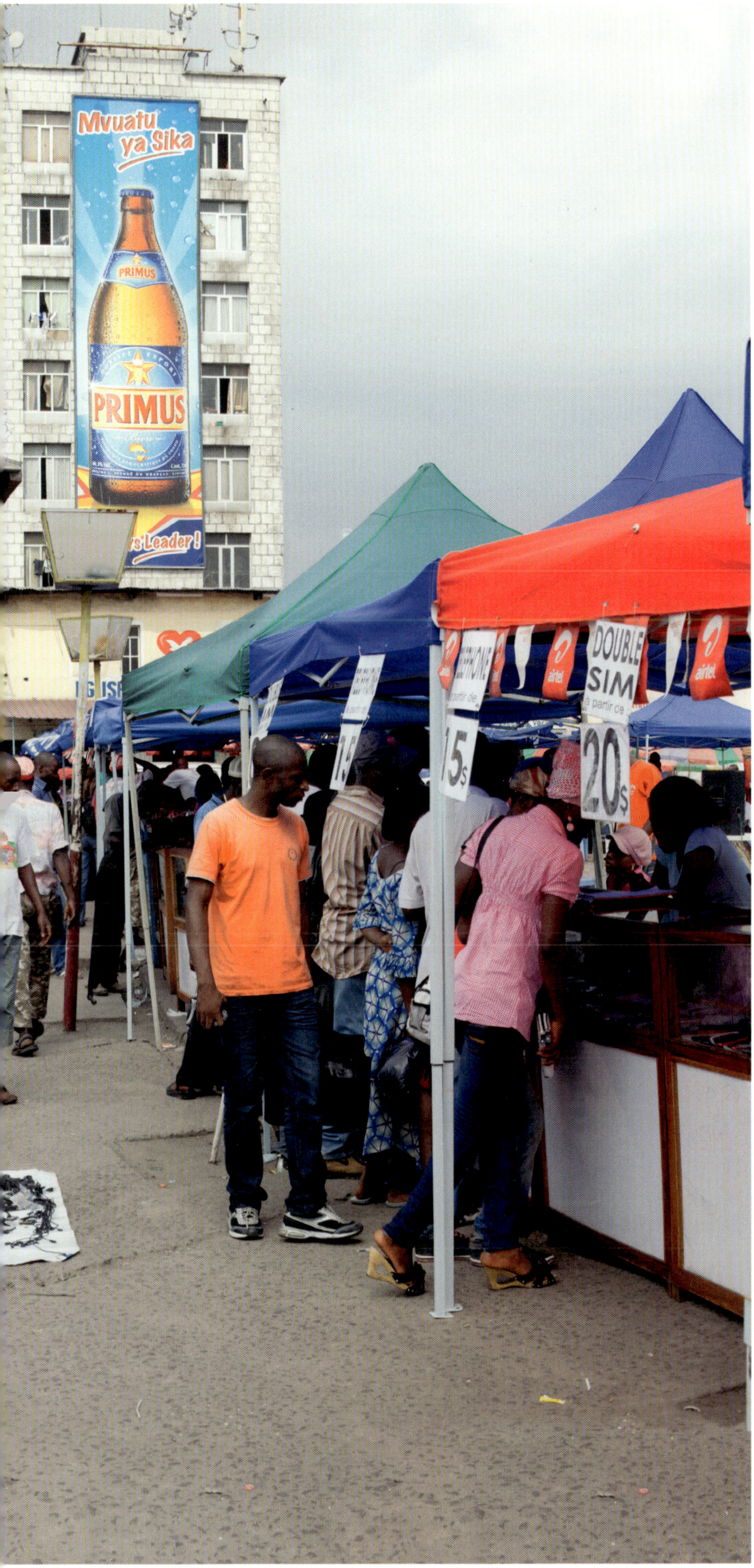

PLACE DE LA VICTOIRE
KINSHASA

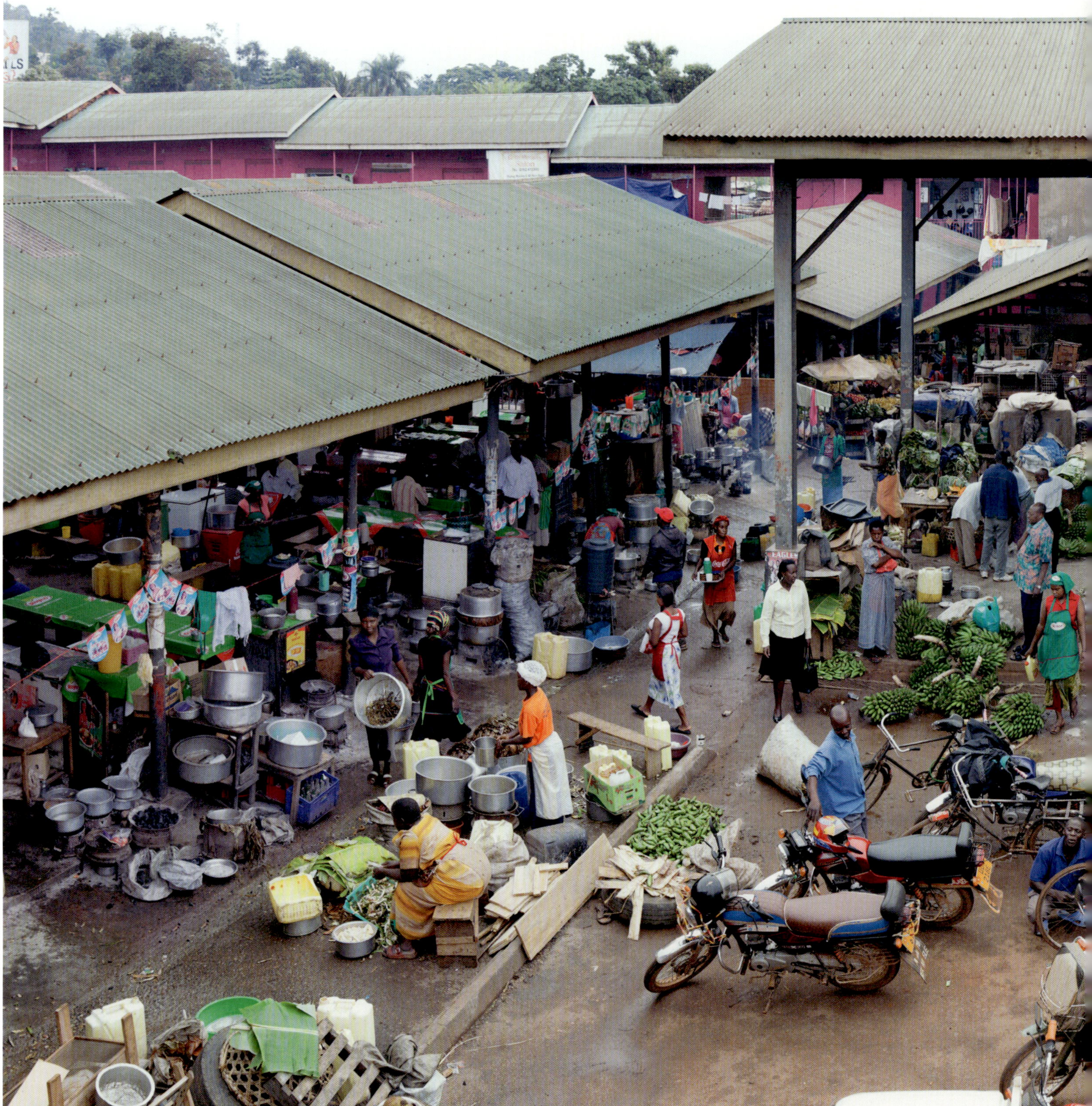

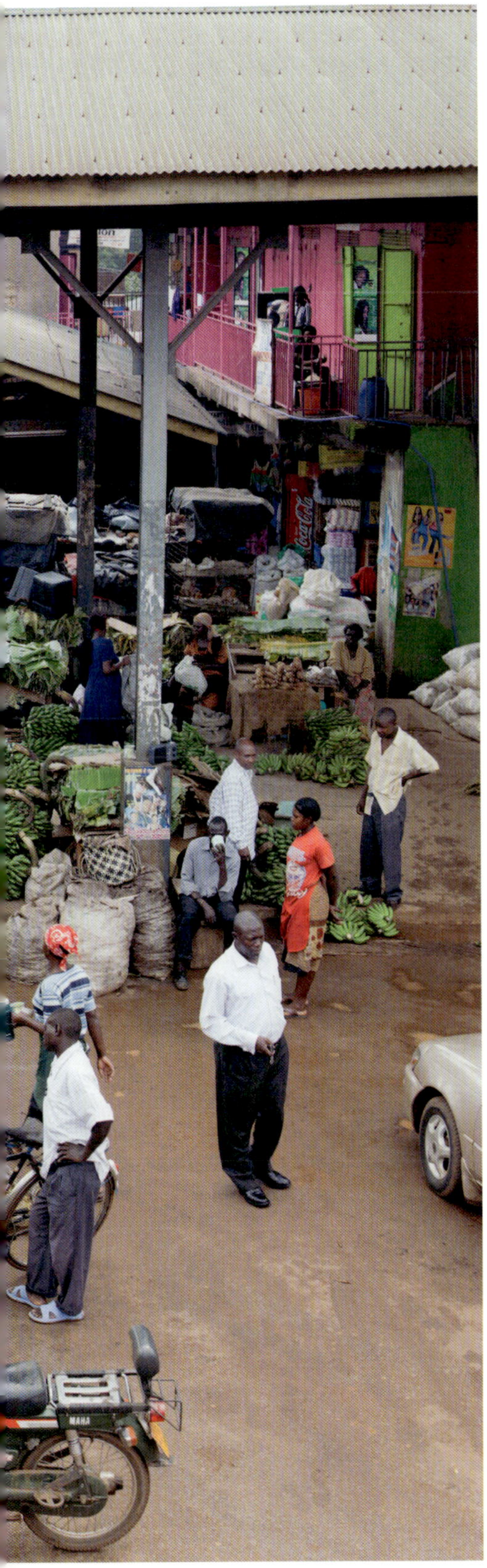

BUGOLOGI MARKET
KAMPALA

PONTE DE VILLA DU GAMEK
LUANDA

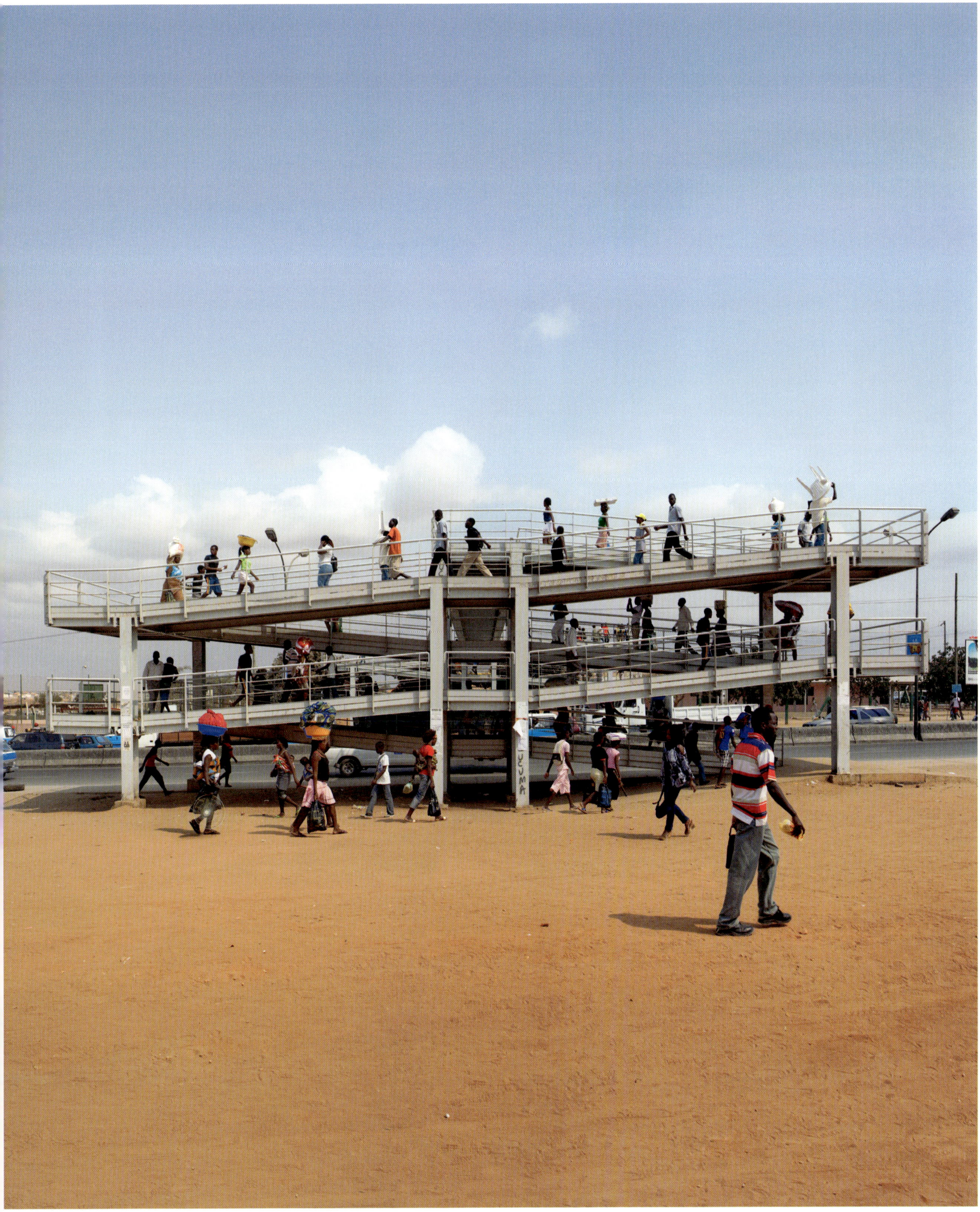

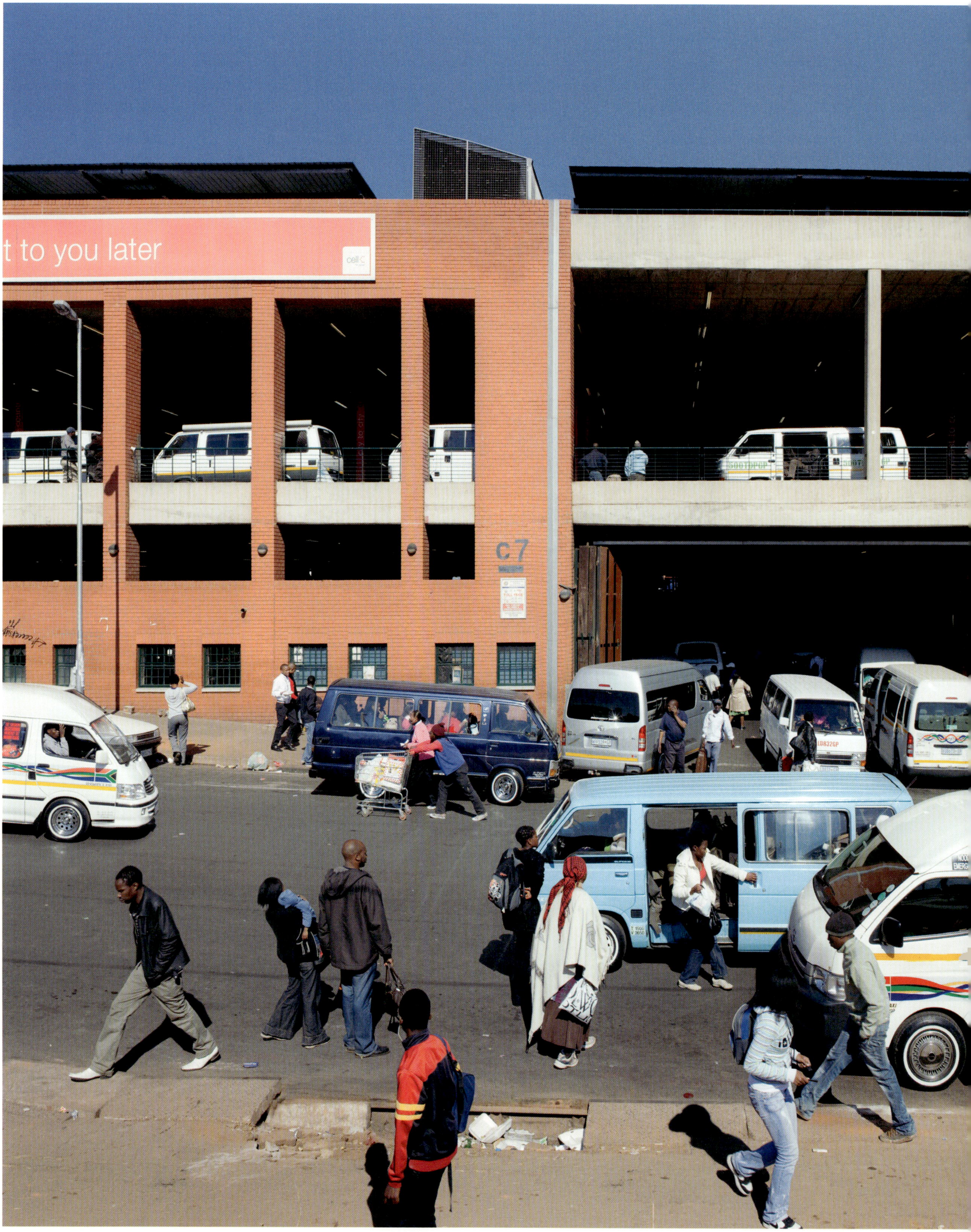
t to you later
C7

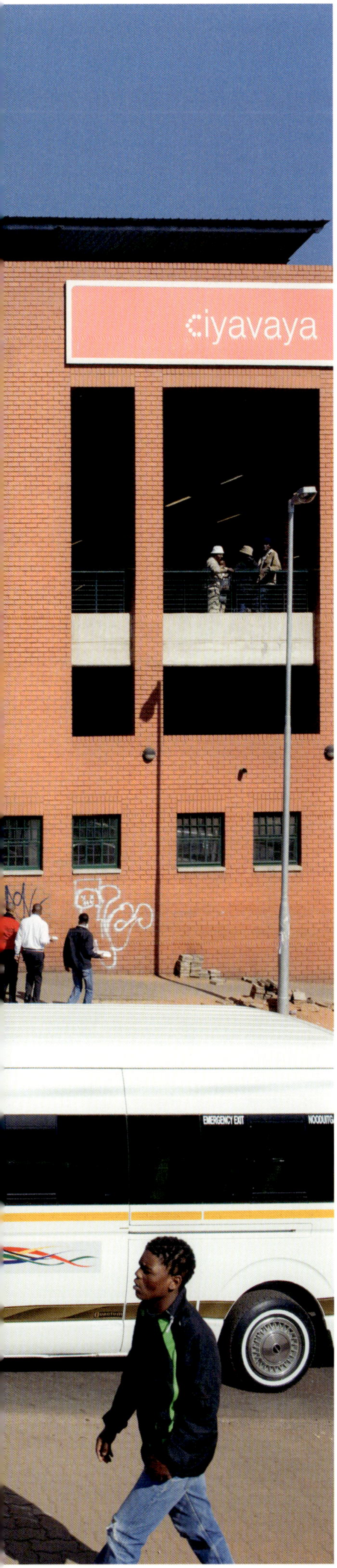

GWIGWI MRWEBI STREET (METRO MALL)
JOHANNESBURG

TSUTSUMAN, ALEXANDRA
JOHANNESBURG

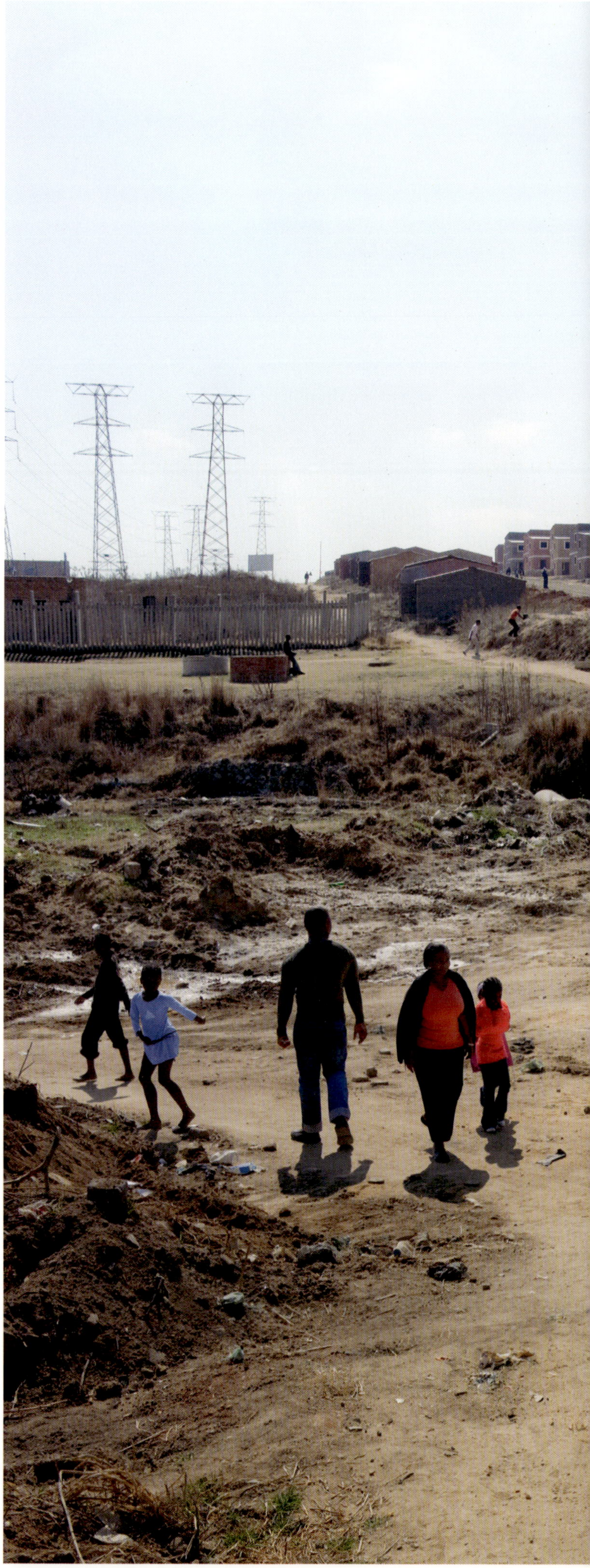

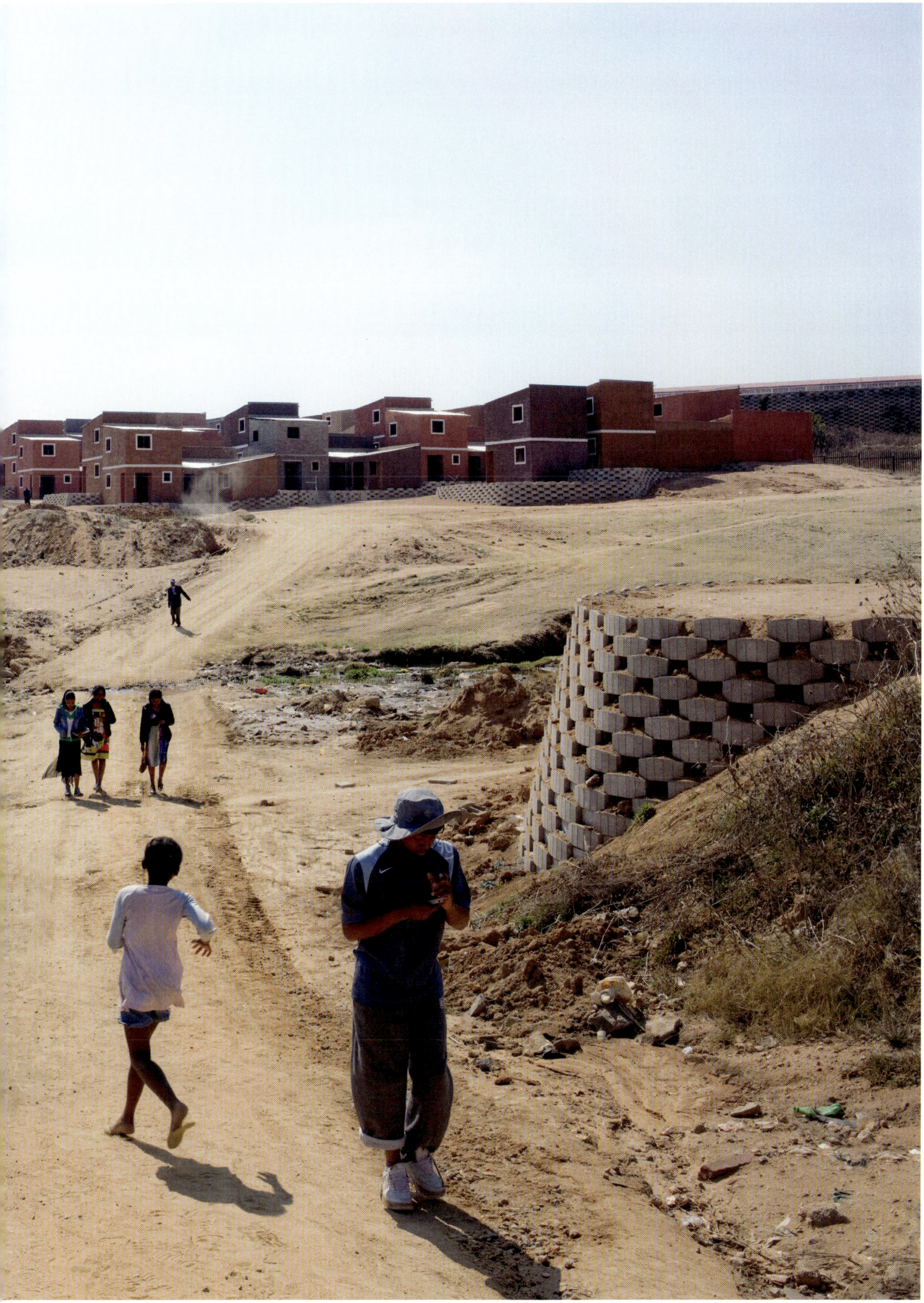

This City

CHRIS ABANI

1.

This city is as much emotion as it is reason. As much magic as it is fact. It is as much present and undiminished as it is lost to the shifting sands of the lower Sahara. It is as much African as it is possessed by a fevered European imagination. This is what African cities are—eternal ideas that persist even when the living have turned their desire to somewhere else. The endless performance of hope in the face of overwhelming odds is what compels the modern African city to continue, to resist anomy.

2.

There seems to be something inherently nomadic in human nature: a need to migrate, to move, to track our existence in motion. We expand over continents, cultures, and of course, through time. Or is it perhaps the very notion of time itself as something that is always moving that we seek to embrace? Every moment is already lost by the time it is realized. So we are always playing catch-up with ourselves, everything becomes understood in a perpetual hindsight.

So it seems the city itself is the impulse to revisit and simultaneously contain nomadic tendencies. To arrest the drift of time in a form that appears to be static but is in fact in constant motion. Cities are as frustratingly inconsistent and paradoxical as humans are.

3.

The city is peripatetic, made up of a fast churn of people.

4.

Echolalia: the automatic repetition of vocalizations made by another person. It is part of the echo phenomena most closely related to echopraxia; which is the automatic repetition of movements made by others.

5.

Roland Barthes describes photographs as unclassifiable. The same can be said for the city. Maybe because of this medium's inherent capacity to take on form, to evolve, to be altered, to be transposed, transported, and to enact and simultaneously codify time and subject. And yet, like the intricate prayers and symbols etched onto a Tuareg ring, they remain peripatetic, a virus, and a simulacrum that stutters and repeats itself endlessly. Making the moment larger than itself, beyond even what it could imagine, a kind of quantum propagation of itself. Suggesting the idea that an event can repeat itself constantly, becoming an echo of an echo. Echolalia. This is what an African city is: an unclassifiable, transitional echolalia. It is unlike anything ever considered in Europe, replicating itself endlessly, like a fractal.

The city is a living photograph.

6.

Here, as in Angola, are gleaming neighborhoods and apartment high-rises that have never been occupied. Built by Chinese investors, the houses and apartment blocks sit on well-manicured grounds with an internal network of roads that are in better condition than those in the capital. To see this is to see the simulacrum of a ghost, a ghost of a thing not yet born. A real dilemma.

7.

To step into this city is to cross borders, because with the permeability of African borders, their recent articulation during colonialism, and the fact that they often conjoin or separate existing ethnic nation states, is to embark on a transnational project. And because of this key quality, transnationalism, to photograph the African city is in effect to photograph every city. All cities can be said to render one city, through time and place, from Timbuktu under Mansa Musa in the sixteenth century to modern Johannesburg. The trace is there, the echoes are loud.

8.

In the West the continuous pursuit of "authentic" cultures elsewhere is puzzling. More puzzling is the refutation of a technologically up-to-date culture in Africa as real. Instead, the more rural or mired in the past the moment, the more "authentic" it is considered. To see Tibetans working the fields with medieval tools is considered "real" or "true" Tibetan culture. That would be the same as if an African came to the United States seeking out the difficult agrarian life of early settlers as a more "authentic" or "true" American culture, and ignoring this current moment.

This is of course based in the biased view that only the West evolves and that the actual current state of Westernization is not itself a product of a larger global conversation going back as early as the eighteenth century. For instance, the word shampoo and the product related to it come to English and Western culture from Hindu culture under the British Raj.

9.

Transience is a word that best captures the spirit of an African city, oddly enough. We tend to think of cities as fixed loci, monolithic statements of a rooted and affluent culture. But it feels like cities are more closely linked to Bedouin/Berber nomadic settlements. Why would I think of cities in this way, as entities always in motion, more peripatetic than fixed? Well, it's because true cities are always evolving, always being rebuilt, reconfigured, reassigned, buildings and other structures constantly shifting like sand dunes.

There are differences between cities of the new order and cities of the old order. Old order cities have, if anything, devolved into towns almost, in their consciousness, if not in their design and built environment. Becoming so self-referential as to have nothing new to say, becoming puns of themselves. Cities like New York and London and even, say, Jerusalem, fall easily into this group. New order cities are always changing, always adapting, not just structurally but also in consciousness. These cities are in dialogue with other cities and cultures, always teetering on the very verge of collapse, implosion, always hovering on the apocalyptic. They are places where motion is ceaseless, where the structures are schizophrenic, where the people are a constantly changing composite of race, culture, and possibility. Modern cities are chaotic beings, always perched on the edge of apocalyptic ruin, be it from tsunamis, tornadoes, earthquakes, sand-storms, and a shortage of land in relation to an excess of human explosion. These cities are Lagos, Mexico City, Dubai, Beijing, and Manila.

The city in Africa, as a concept and as a place, clearly falls into the category of newness, of change, of nomadic pursuit, of constant flux. Either from man-made or natural challenges, or both. Shantytowns and ghettos buck up against some of the most architecturally diverse and affluent neighborhoods in the world.

10.

Space and place bleed into one another in this city. Some locations we think of as place, because place always implies an inherent story, an actual consciousness that has developed over time and repetition. Space is however, neutral, or as neutral as one can get in this pattern and narrative loving ape brain of ours. Space is necessary for all evolving cities; it is the moment of slippage, of random regeneration. Sidewalks in an African city are such places—they support a varied if impermanent ecosystem—shops, cafes, work places, barber stands, and more get set up and stripped down every day.

These areas of slippage make up the invisible cities-within-cities, like the psychic fault line in the fabric of Los Angeles called Alvarado Street. One end rises in the Hollywood Hills and dances through poor neighborhoods, rich neighborhoods, through areas where you can buy fake ID's to MacArthur Park off Wilshire, itself part of the city's subconscious.

11.

This city can only be photographed in motion, the blur of movement the only thing that slows it down enough to bring it into visibility.

12.

Alleys. The Yoruba call them *Koro.*

13.

As in the alleyways in Ojuelegba, Lagos, perhaps the busiest and most famous crossroads in Africa, an impromptu lean-to grows to house an impromptu night school. One wall is the massive concrete pylon that holds up the freeway above, the other a bar in the infamous red light district flourishing under it.

14.

With all the roar and dust of the African city, the way fecund vegetation reclaims anything left unattended too long, one cannot be fooled into thinking that anyone can really trust a city. It is at best a fragile negotiation with nature, with the elements. The Yoruba of West Africa have known this longer than most—they began living in the urban clusters that would forge modern cities over seven thousand years ago.

15.

The city is a form that resists singularity of time and place. There are ways in which cities take on the emblematic nature of poetry—a series of disembodied simulacra that become like viruses, constantly changing and evolving. Enacting always the event of their capture but also all the different ways they can be approached, read, classified, recontextualized, and reimagined. Like all good art, cities guard as many secrets as they reveal, occupying both the intimacy of the snapshot and the archival importance of documentary.

I used to go to many of the flea markets in London—Greenwich, Camden, Hackney, and even the Thames South Bank Promenade—and rifle through boxes of old photographs. I like the idea that they had traveled far to find their way into these boxes. Displaced as they were from context, meaning, and ownership they became archeological scraps of lives that never seemed to stack up too much in themselves, in their lived time and place, but with this scattering, they took on new historical lexicons in the artistic scrapbook of my mind. Or perhaps these were all the romantic stirrings of a mind addled by Earl Grey tea and the soft sun of England's summers.

This city is that box, the disheveled, displaced lives that end up here, so many snapshots of humanity, always hinting at, but never quite capturing the whole picture, the whole story.

16.

The colonial dream of order, of European order, is thwarted by the African city. In Western cities, planners try to instill an order over chaos, carefully staging the architectural—grid lines, streets and the cultural, sleek clothing, business looks—all aligned carefully, all of it to repress an old fear of the chaotic. But even in the West, cities resist order, leaning always towards chaos—the errant clump of grass growing halfway up a brownstone's wall.

It is not that African cities are not staged, they most certainly are. It is simply that there is an understanding that the staging is so complex it cannot be held to one look, one consciousness. It is a staging that takes into account uncertainty, the swirl towards chaos. If you drive in an African city you might find it hard to understand why there are not more accidents. How this elaborate chaotic dance seems to follow a fluid algorithm lies in the acceptance of the unpredictability of human behavior. At its most instinctual African cities seem to accept the architectural as the hardware of the city and people as the software.

In a Western city the court building will mostly be dominated by lawyers and other users of the court—all dressed appropriately. In an African city, the court steps hold lawyers and petitioners, but also beggars, food sellers, people offering typing services, and vendors selling everything from videos to phone cards. So to attempt to stage a city here would automatically fail.

17.

It is the simple ordinariness of the scenes, the sheer quotidianess of the subjects that transform these cities.

18.

To witness this city, or any city, at 4 a.m. is to witness silence. You are stopped at a red light that scrutinizes you. There are no other cars in sight, no people. Just the shuffling hulk of darkened buildings, façades inscrutable, hunkered down for the night. And also the lights that beckon ceaselessly to phantom traffic. It is as though someone threw a switch somewhere and half of the spirit of the city went to sleep. The flashes of traffic lights are like neurons dying slowly in a brain. That's when you realize it. The architectural elements are not the city. They are simply material anchored to a specific site on a specific night in April while you wait for this specific light to change.

A real city comes alive when people flood its veins and arteries like so many ants in a complex nest. A real city is a musical instrument. It is meant to be interactive. To have people sing through its strings and hollowed bones.

Cities are always cellos. Don't ask me why, just accept this fact.

But this city is not its people, either. Otherwise, every Bedouin caravan and every oasis would be a city. But they are not. Without the architectural subconscious to play against and without the people to do the playing, there is no city.

The city is a social entity. A flirt that will hang out at every bar until thrown shamelessly into the street. The light changes. You move forward, unsure why you are suddenly cold.

19.

Cities are the most public of all art forms, available to all (if you have the right papers), interactive, sublime, and ever changing. An exhibit curated by late-night cleaners, bored policemen, and drunks lounging in an empty car park. But we forget the complexity it takes to curate this, the sleight of hand, and instead we take it for granted. But this beautiful, complex and sometimes disjunctive human-architectural backdrop hides as much as it reveals.

20.

It says something that in most European cities prisons, churches, and factories are the three predominant structures that always survive time and decay. They form the triangulation of the cultural state, the trifecta of desires and dilemmas that face the average citizen. There are no enduring structures like this in African cities, even though many, like the Yoruba cities, have been around in some variation of themselves for seven thousand years.

21.

When we think city, we think movement. The frenetic movement of a city terrifies small-town people. They think there is no codification to this movement, no behavioral patterns that are carefully controlled. But we know that movement without codification breeds anarchy, so to enforce codification, cities are designed around the control of resources: light, water, and air.

22.

African cities make the worst lovers. They seduce you with an impossible juxtaposition of tableaus, and when they have you, when you can never leave them or love another city, they break your heart.

23.

This city still wears the vestigial undergarments of its rural past, layered like the petticoats of a Victorian woman—goats wandering down freeways, chickens laying eggs in abandoned cars, kettle hawks hunting rats in the alleys, and an old man so lost he will remain ever lost. And still, what is beautiful remains.

24.

At some point or another cities intersect with imagery. Bas-relief carved into the façades of banks and court rooms, mosques ornately attired with tiles in geometric patterns, cathedrals old and wondrous with interiors that are immune to the noise of traffic, and maps, the cartography that attempts to contain the chaos, arrange it in lines and make it easy to comprehend.

And sometimes, documentation. Photographs that attempt to reproduce the city, to bear witness to the monolithic fact of them, to the essential nature of them, are always subverted. Even in the early twentieth century photographs composed of workers perched atop the concrete behemoths that they are constructing fall prey to the human spirit. For all the grand scale of the Brooklyn Bridge, a lone worker perched like a bird on it reverses the scale and we see the man as the dominatrix in this relationship; that humans control always the city, no matter how ornate a space it is.

25.

People lose their hearts in a city like this, all flat roofs and ochre, like a plate broken carelessly on the side of a cliff. You'll be unable to cipher the clues, the indifferent paths of streets laid out like black canals, no grace, not here. Just crowds, humanity pressed deep and close. Or, the city leaves us like this; lost forever, stopped at a railroad crossing gate that holds everything in a limbo of incredible agony—the train never coming, the gate never lifting, the cars stopped, and a woman selling popcorn from her tray comes back to tap against the window like a fly caught in a screen door.

26.

A minibus painted with green, red, and yellow stripes; a crown of live squawking chickens tied to the roof rack like a headdress of a shaman in dance bears down.

27.

Smoke from a cigarette billows over the heads of women gathered in front of a nightclub.

28.

Under a sign that announces: "Doctor Kareem, Indian-Trained Mystic, Can Cure Aids with Seven Mystical Baths," a young boy sells oranges from a tray.

29.

A young girl in a school uniform makes her way home with a group of friends. They don't speak much to each other, but their shadows intertwine seamlessly.

30.

A man stops under the shade of a tree. He is sitting on the upturned wheelbarrow that he uses to cart goods as a porter. A half bottle of Coca-Cola sits next to him. The newspaper he is reading is upside down. This doesn't deter him.

JUNCTION #2

CITY OF STORIES

“Whether it’s a business district or a garage open to the sky, all the characters display the same sobriety. Neither poor souls in search of civilization or salvation, nor victims of selfish Western globalization . . . They are always anonymous silhouettes moving through life in the certitude that the best is still to come.” *

* N’Goné Fall, “Reinventing Reality,” pp. 7–12

CLOCK TOWER GROUND
KAMPALA

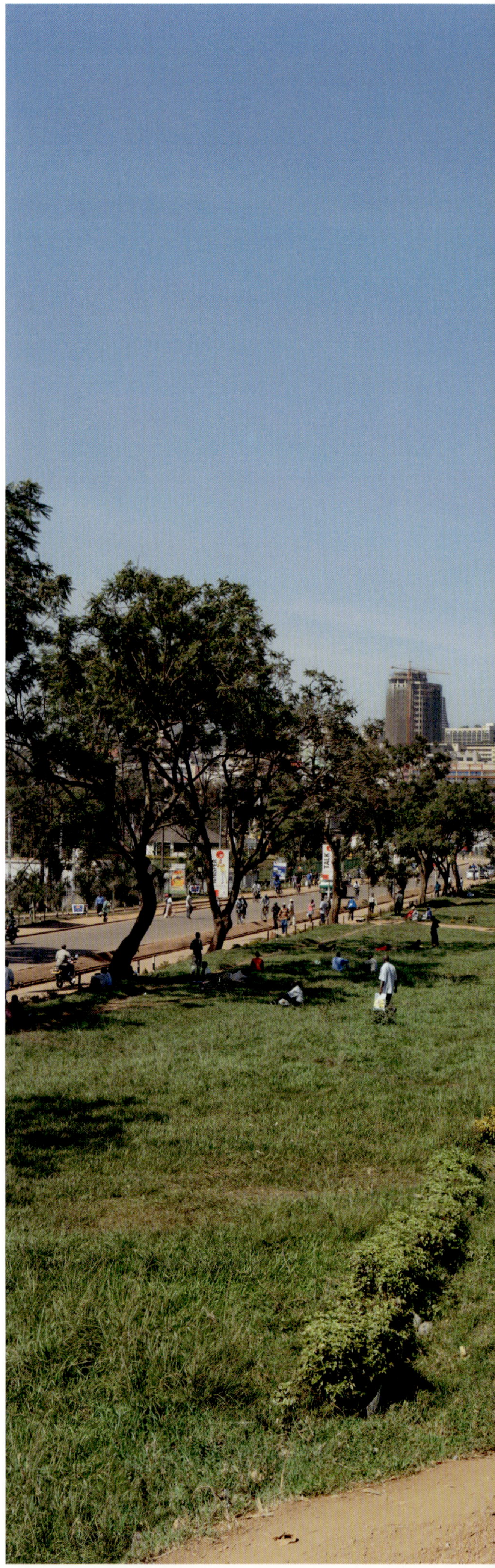

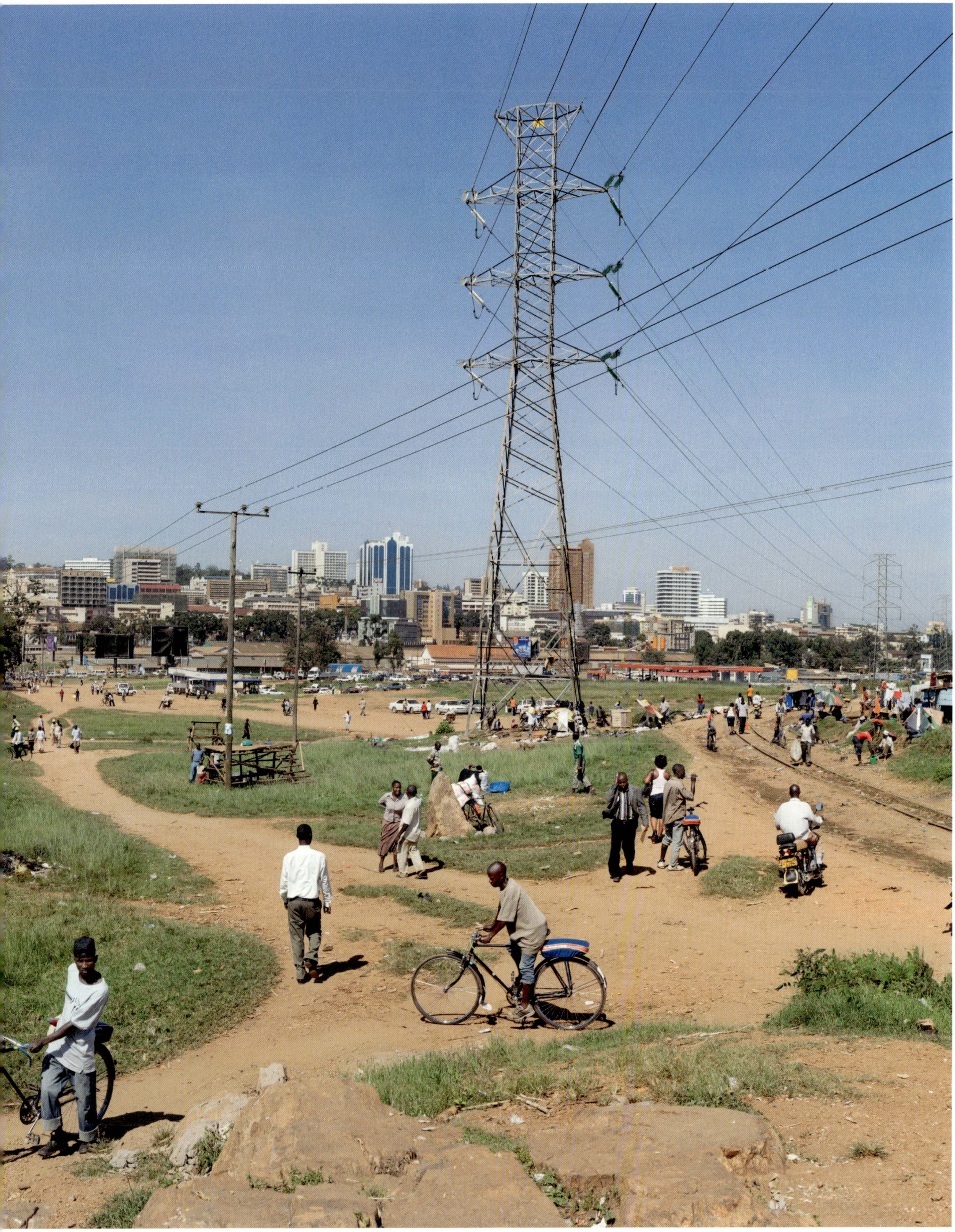

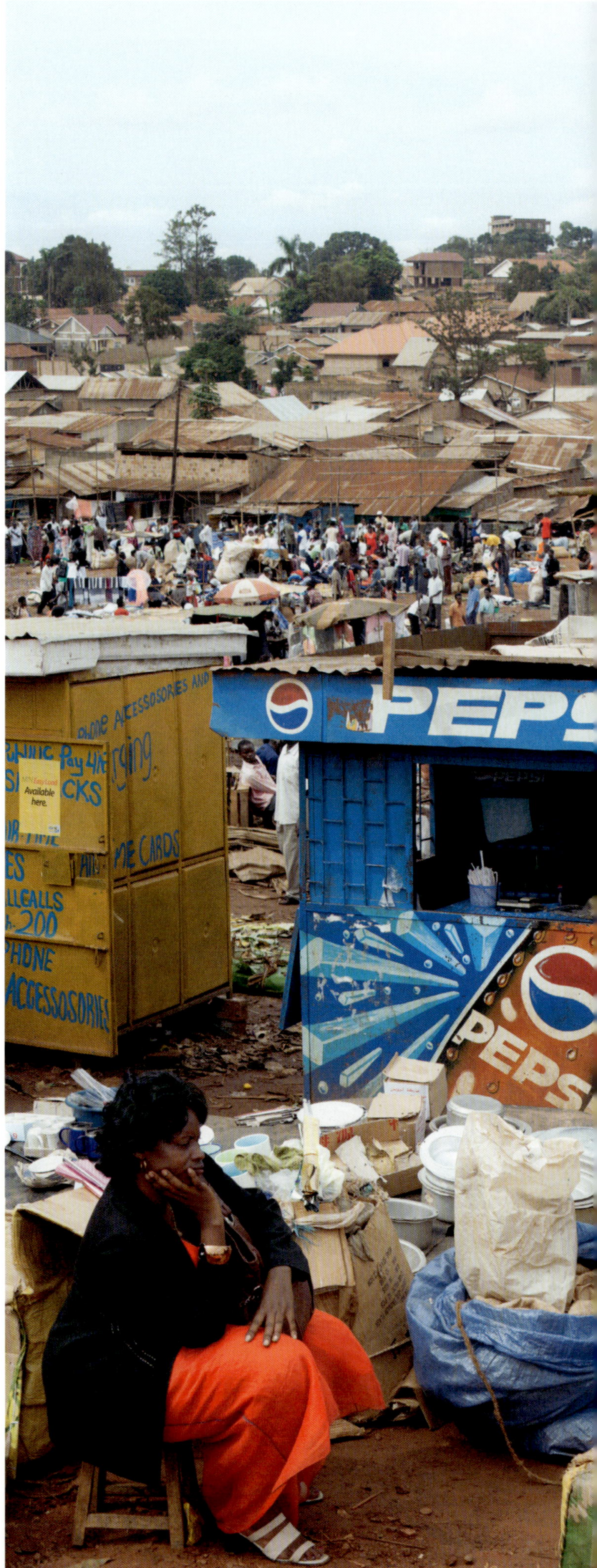

MARKET, BWAISE
KAMPALA

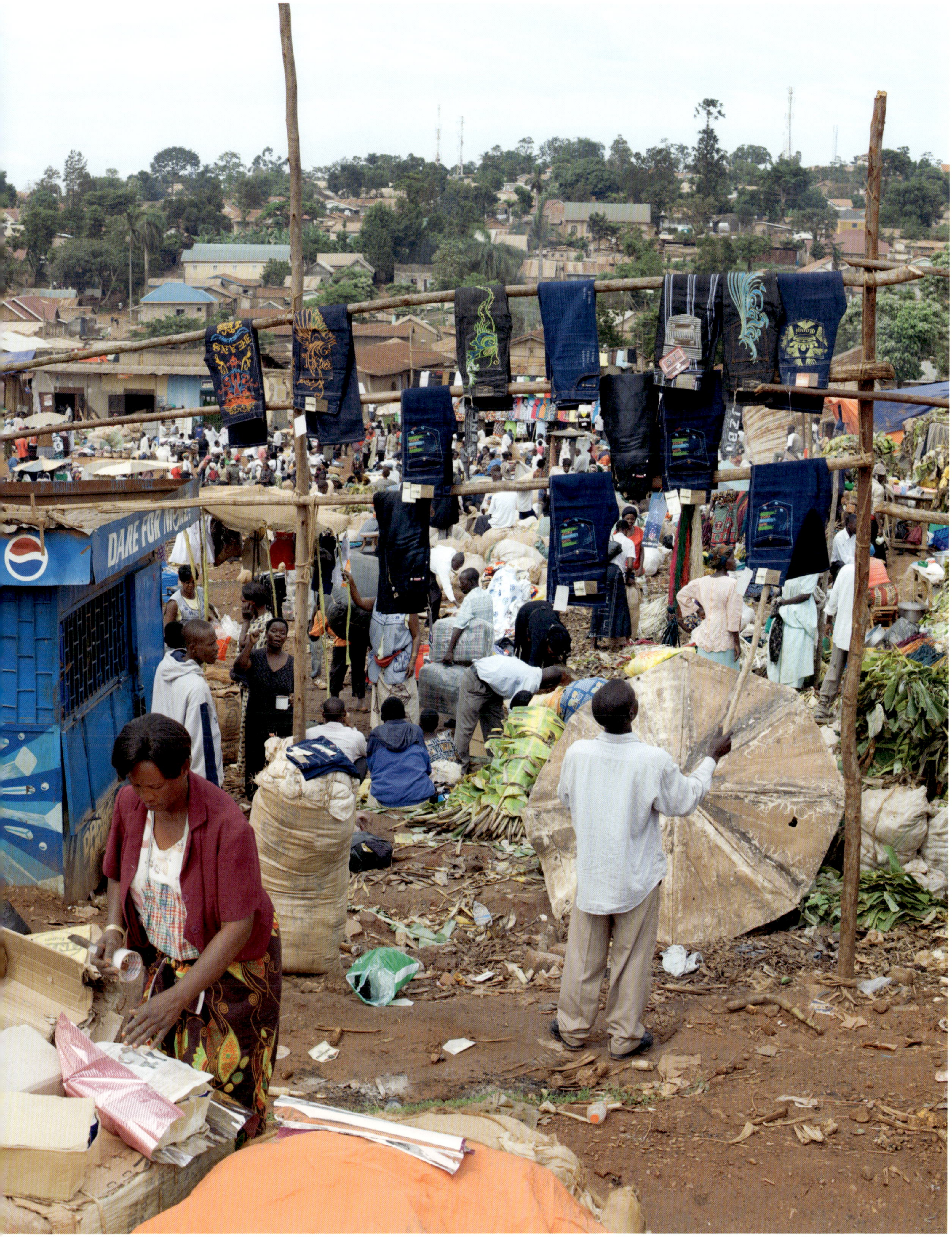
DARE FOR

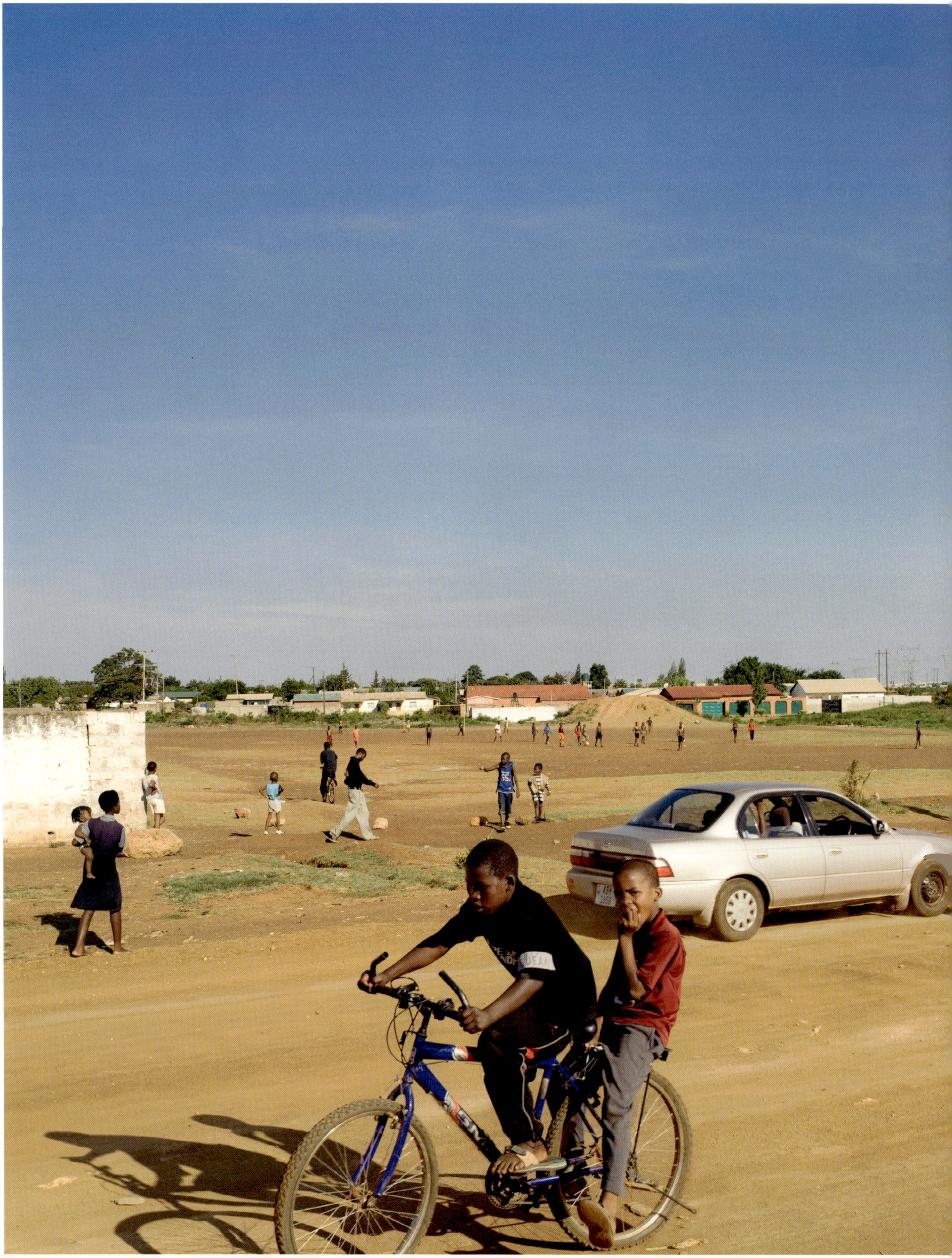

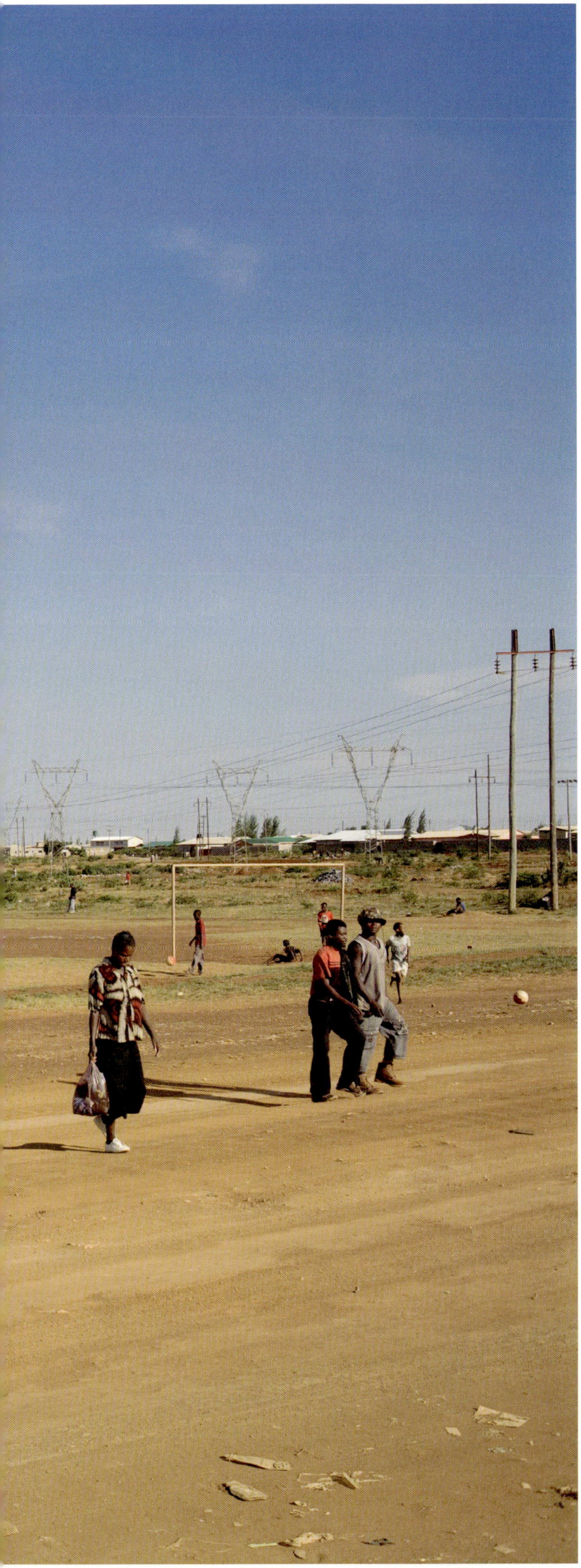

CHILENJE
LUSAKA

ASA BRANCA
LUANDA

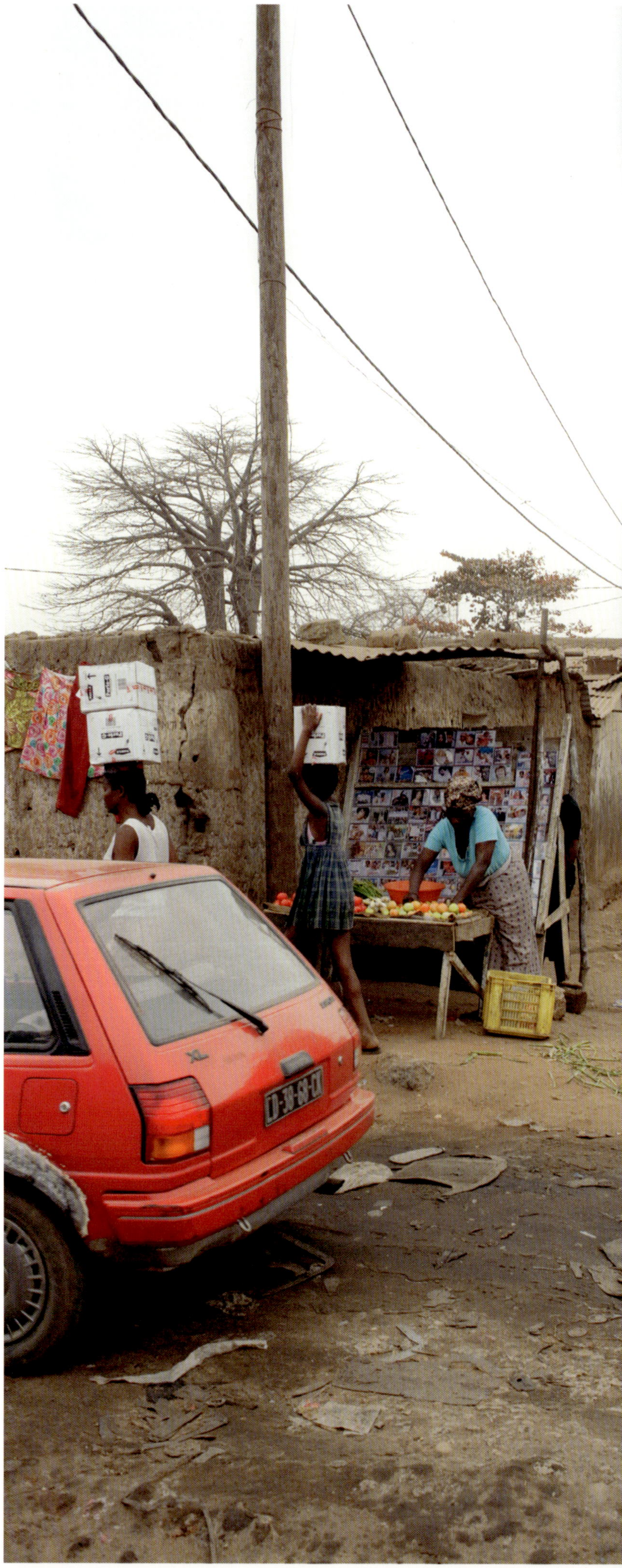

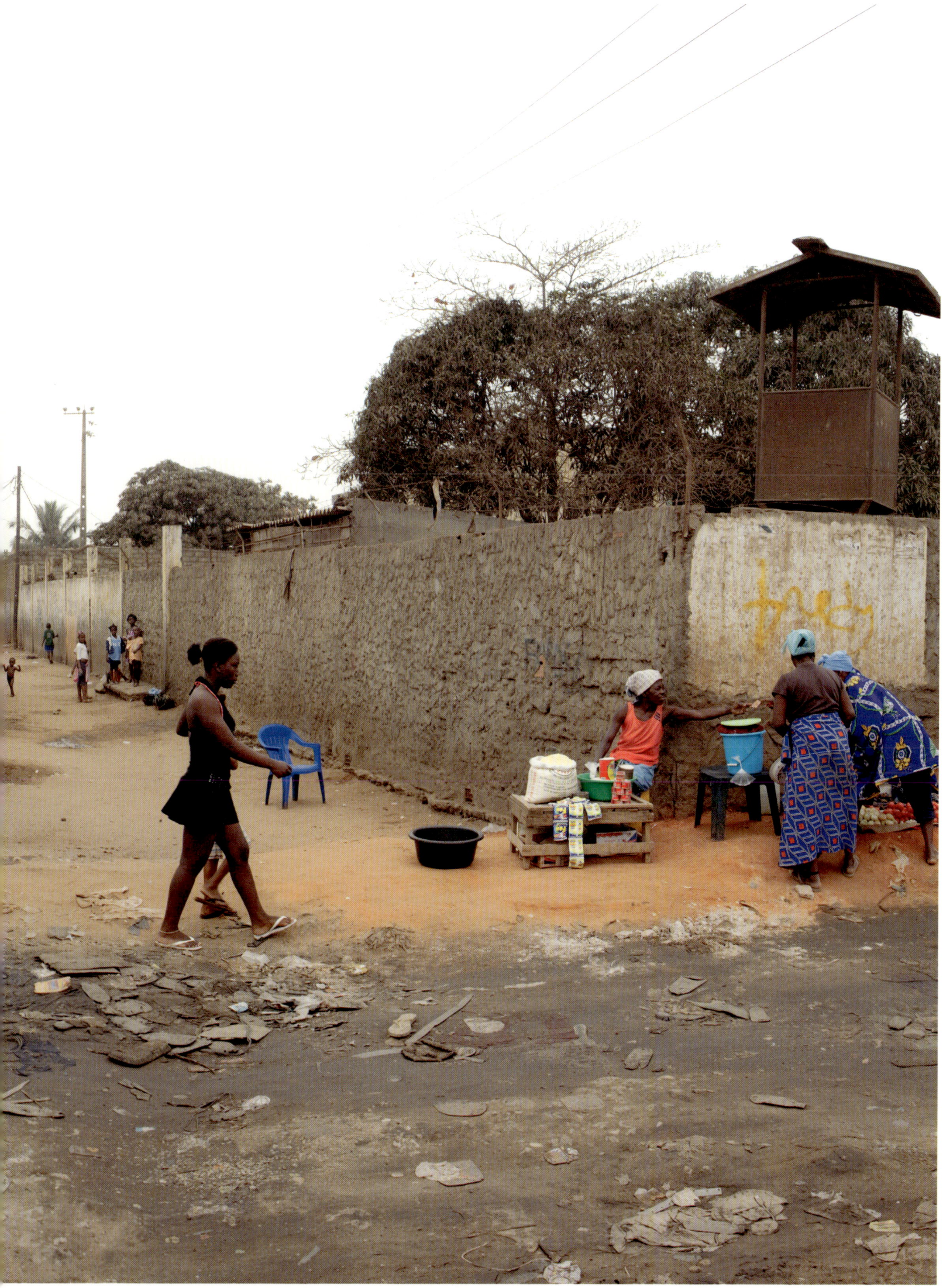

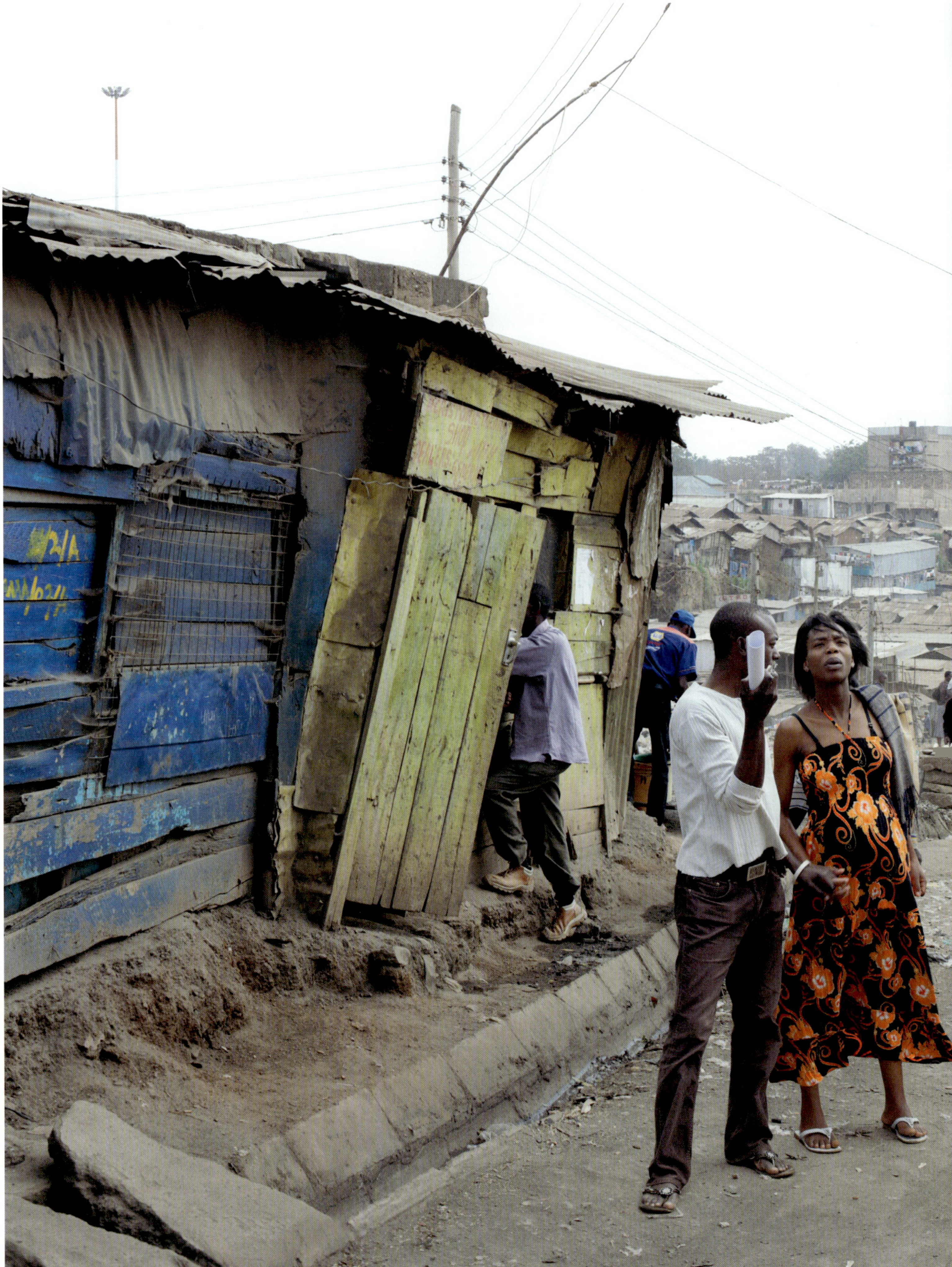

MABATINI, MATHARE
NAIROBI

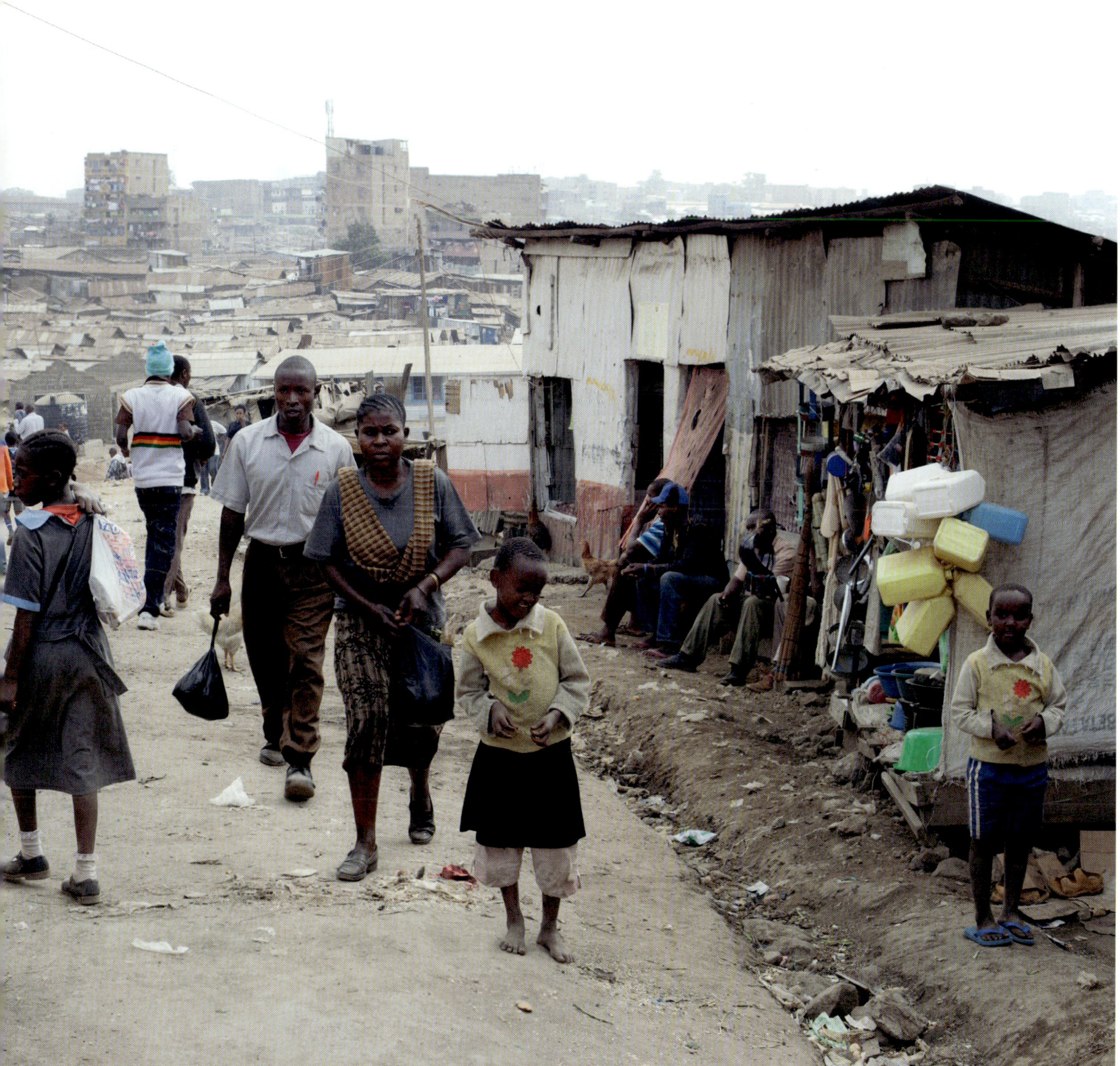

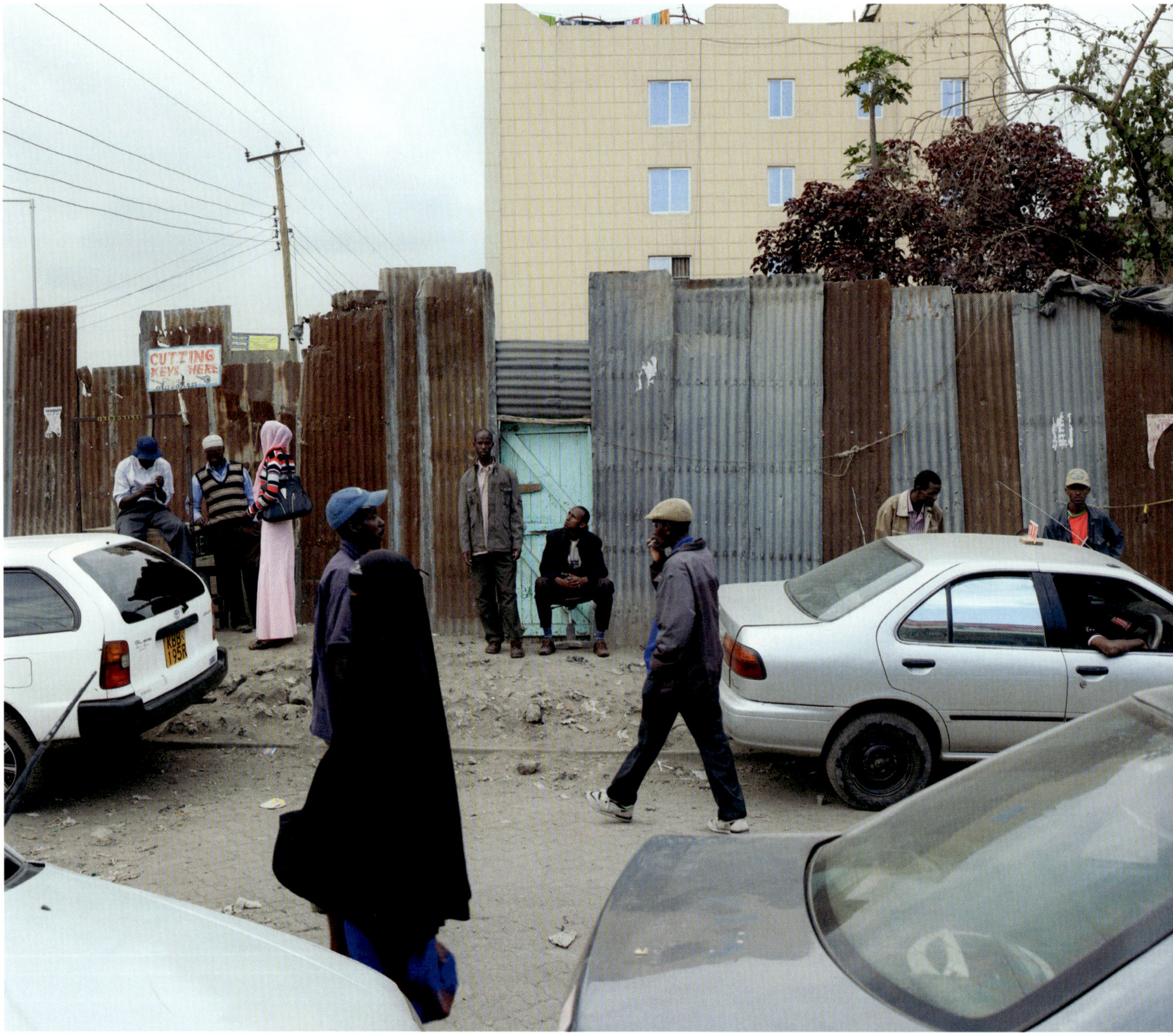

10TH STREET, EASTLEIGH
NAIROBI

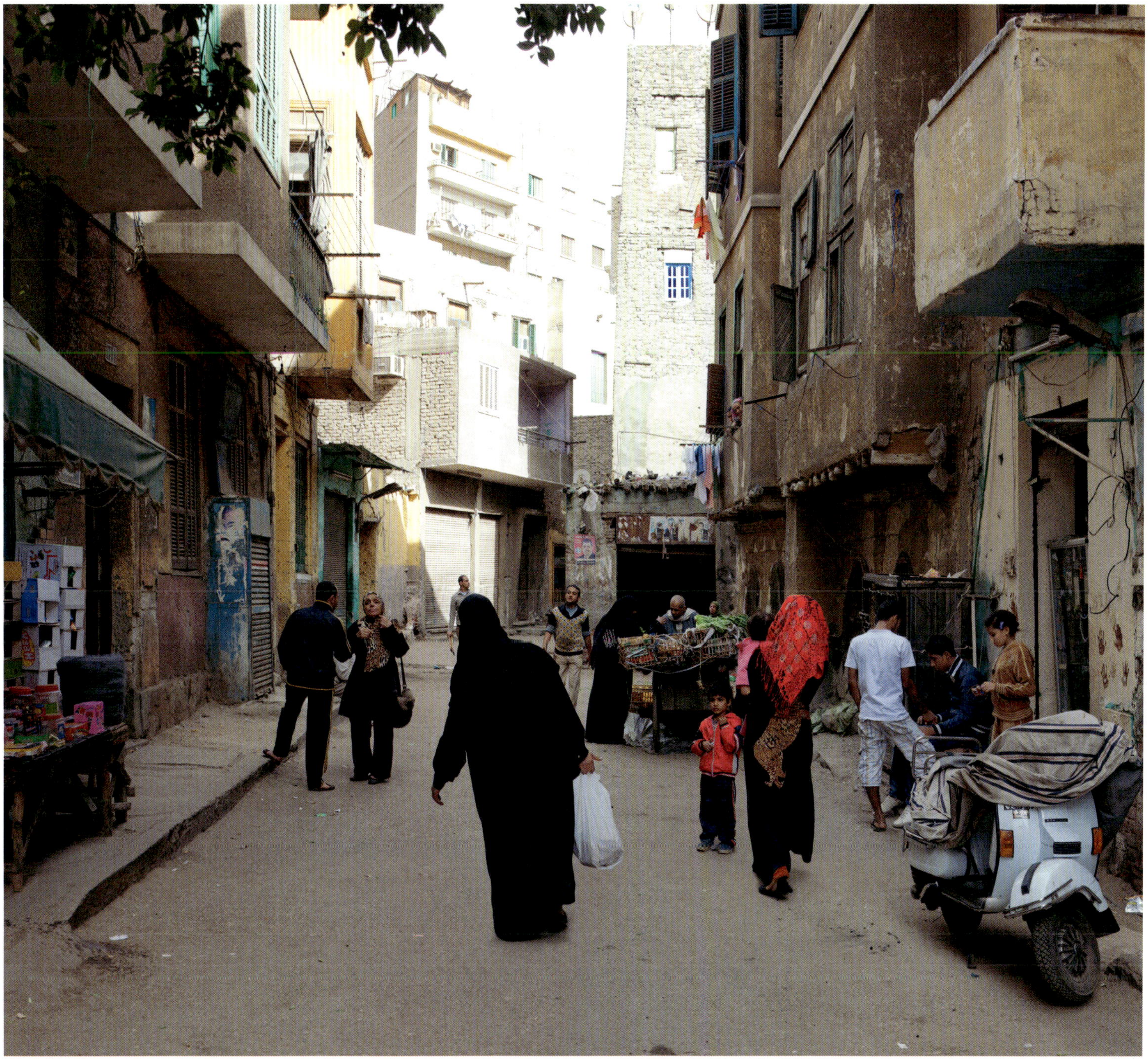

ELSHIKH ALY STREET
CAIRO

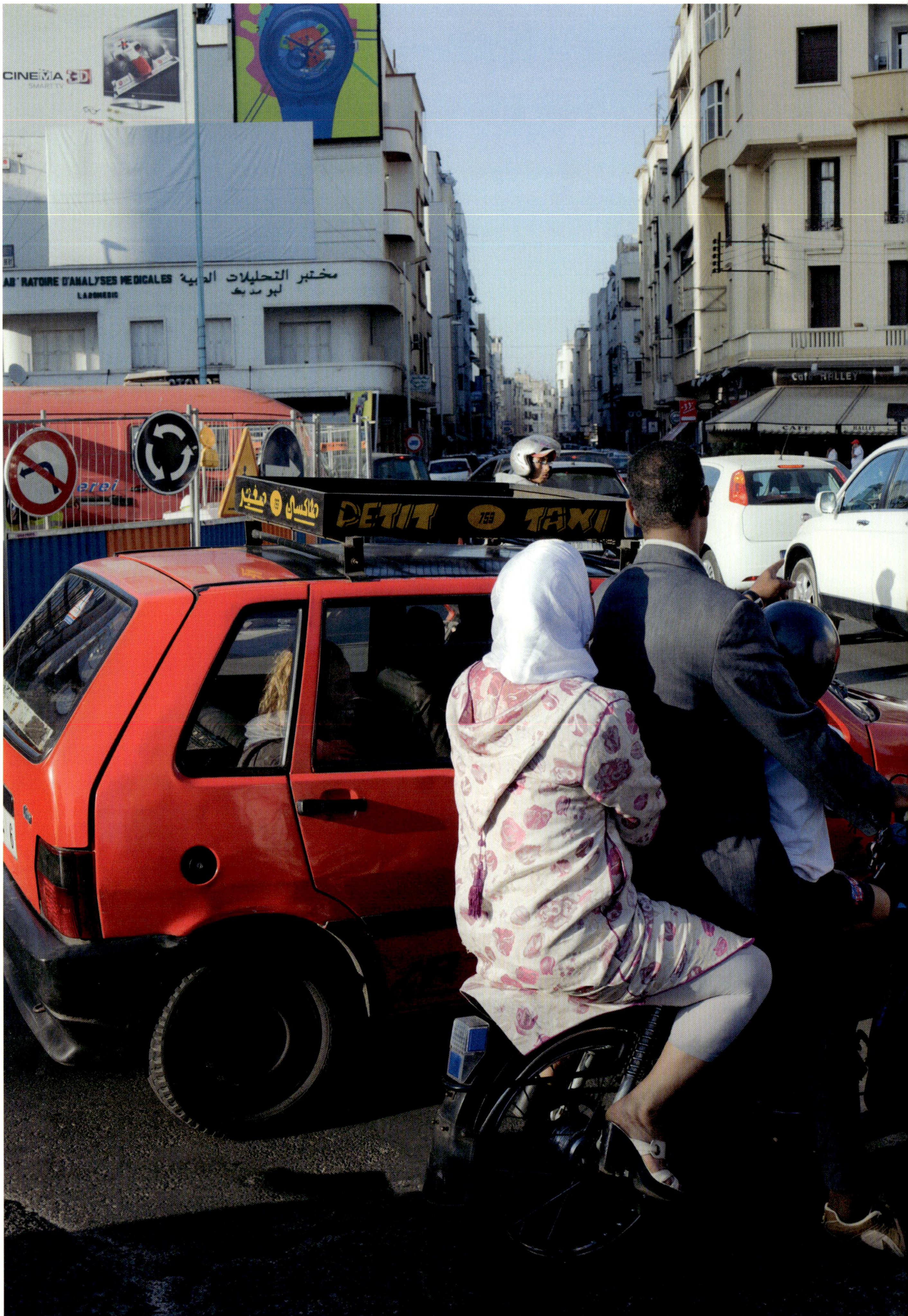

ROND-POINT HASSAN II
CASABLANCA

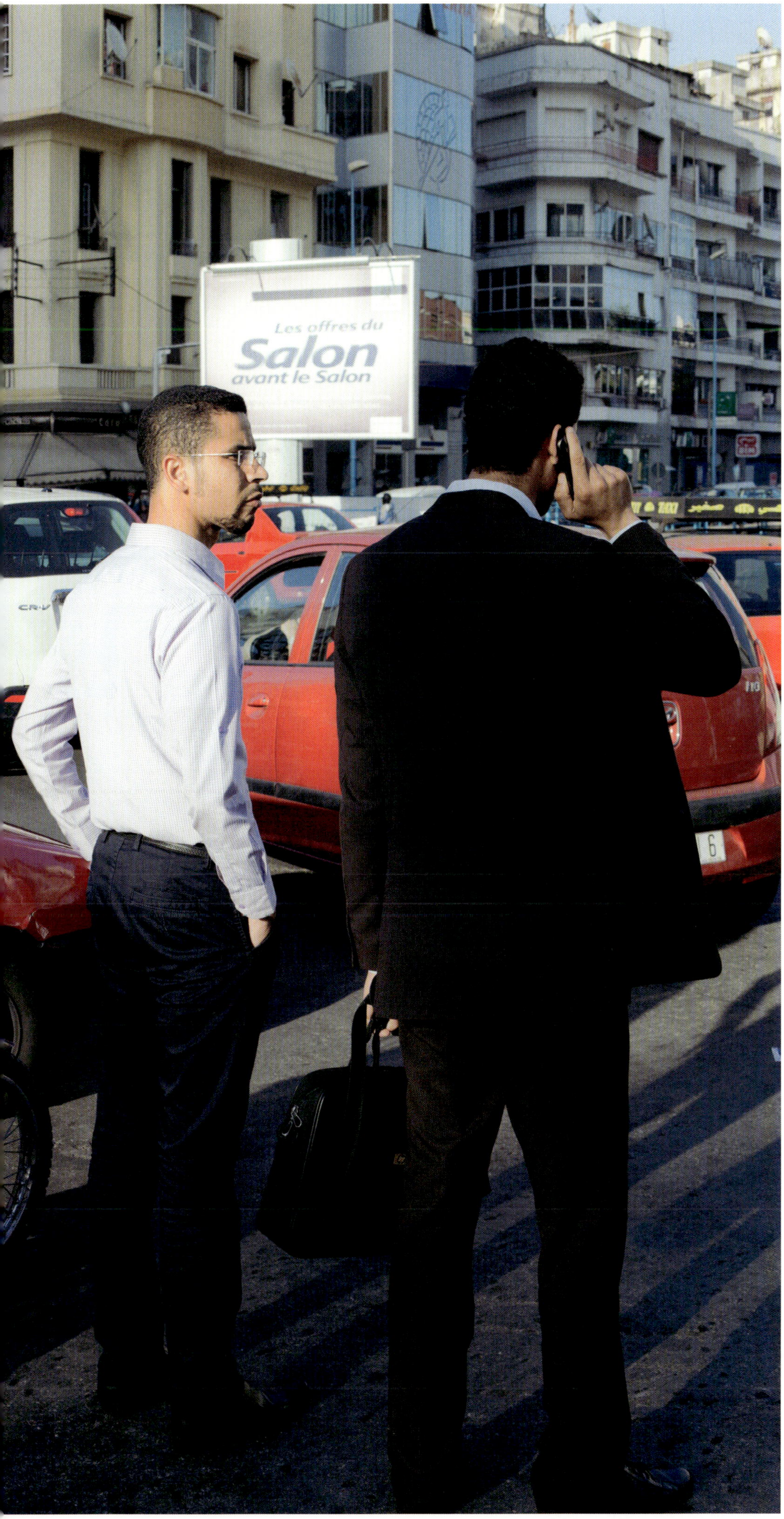
Les offres du
Salon
avant le Salon

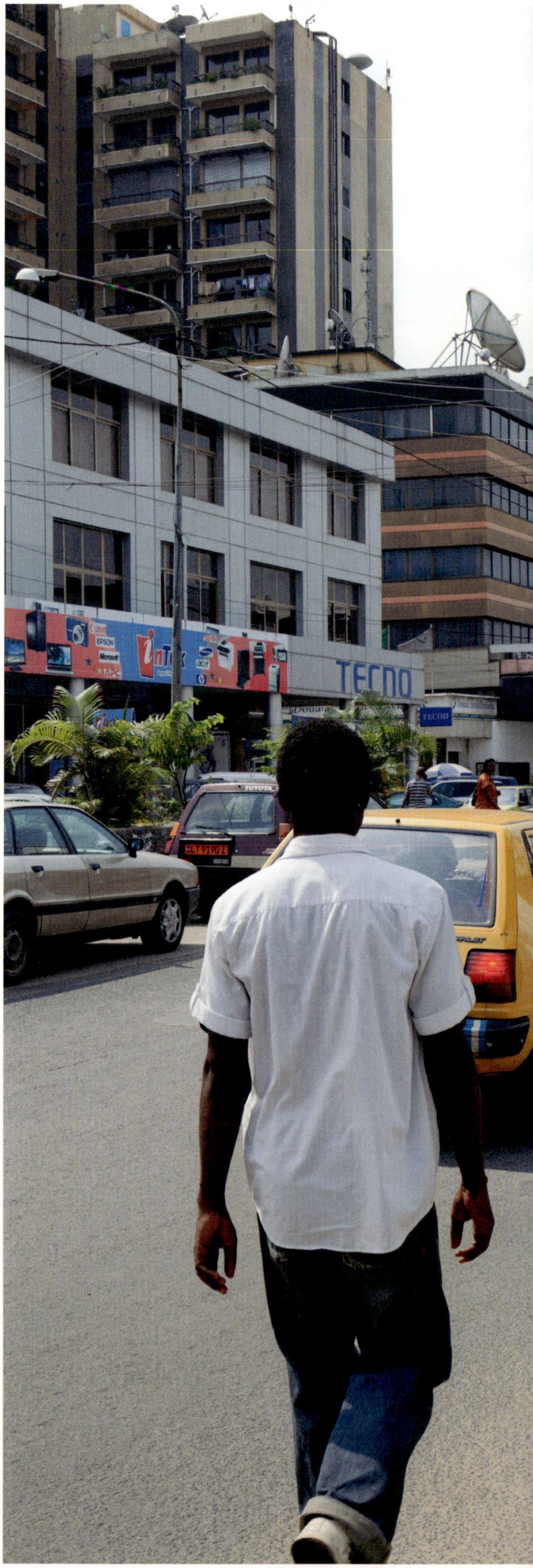

BOULEVARD DE LA LIBERTÉ
DOUALA

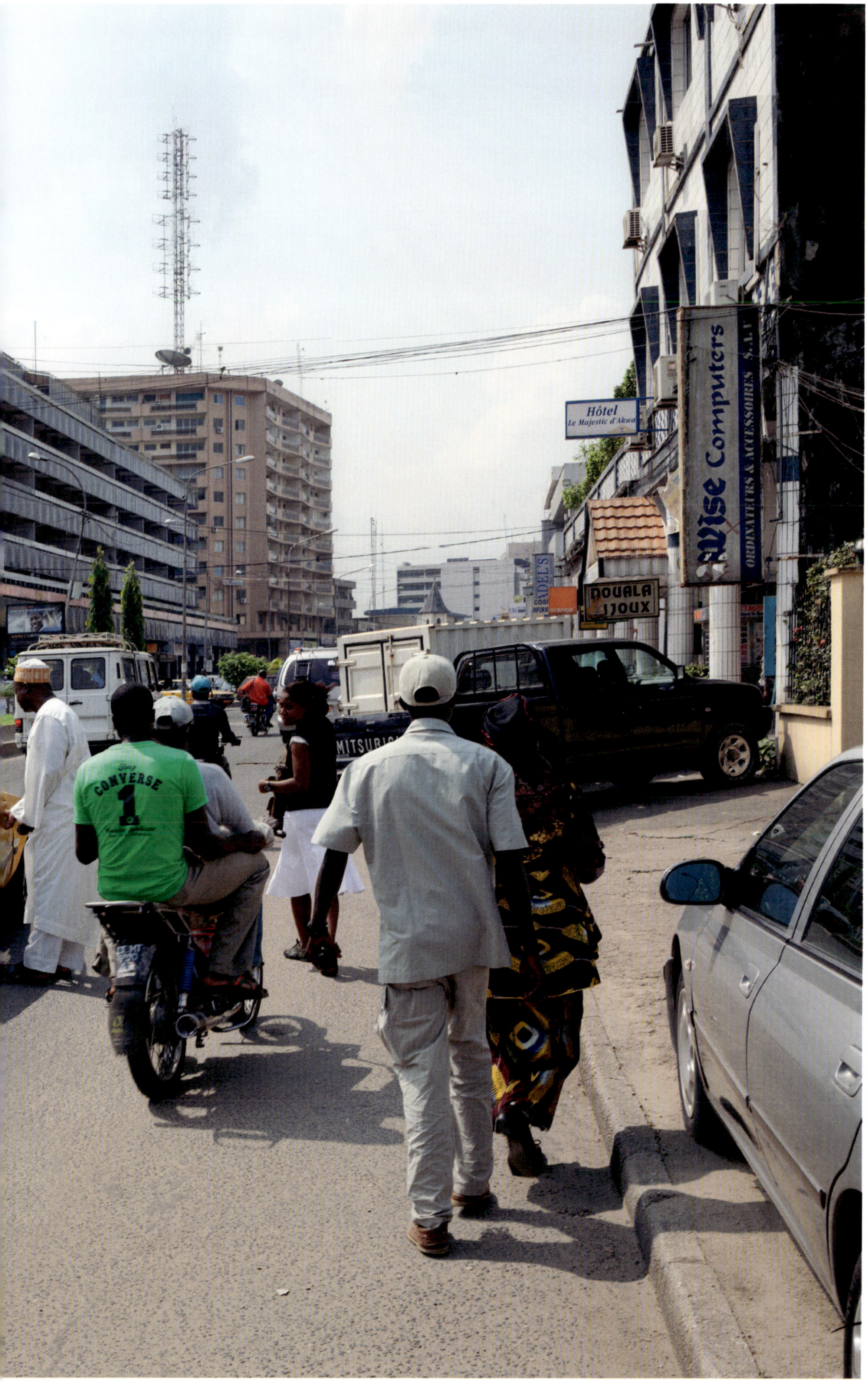

Hôtel
Le Majestic d'Akwa
Wise Computers
ORDINATEURS & ACCESSOIRES S.A.R.L
DOUALA
CONVERSE
1
MITSUBISHI

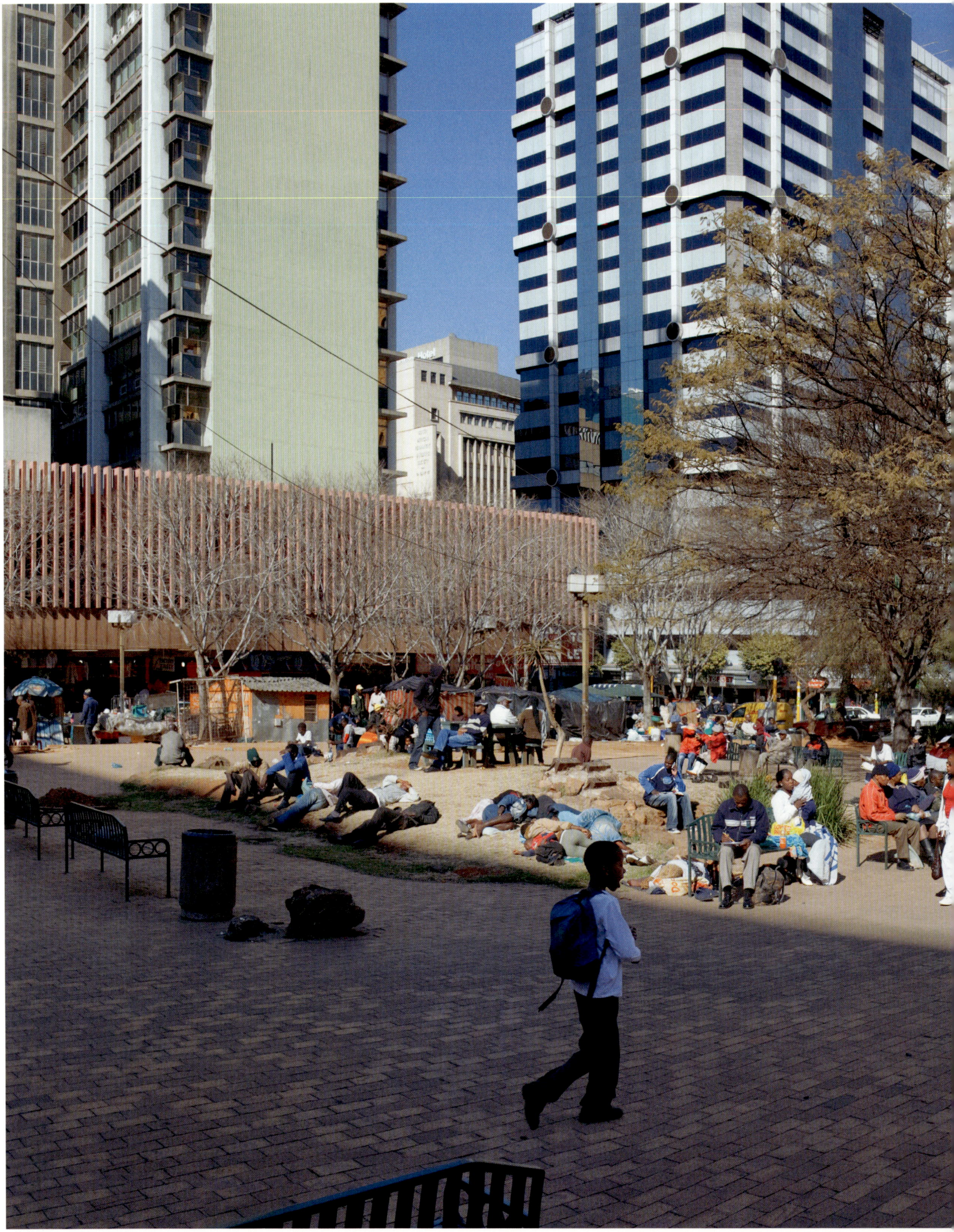

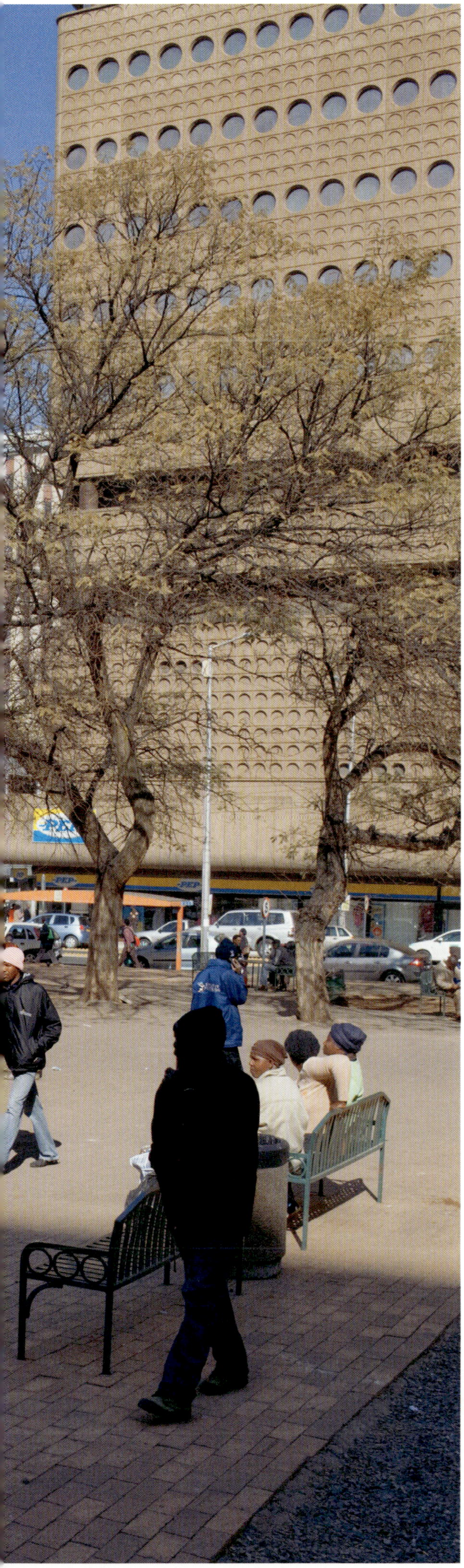

ERNEST OPPENHEIMER PARK
JOHANNESBURG

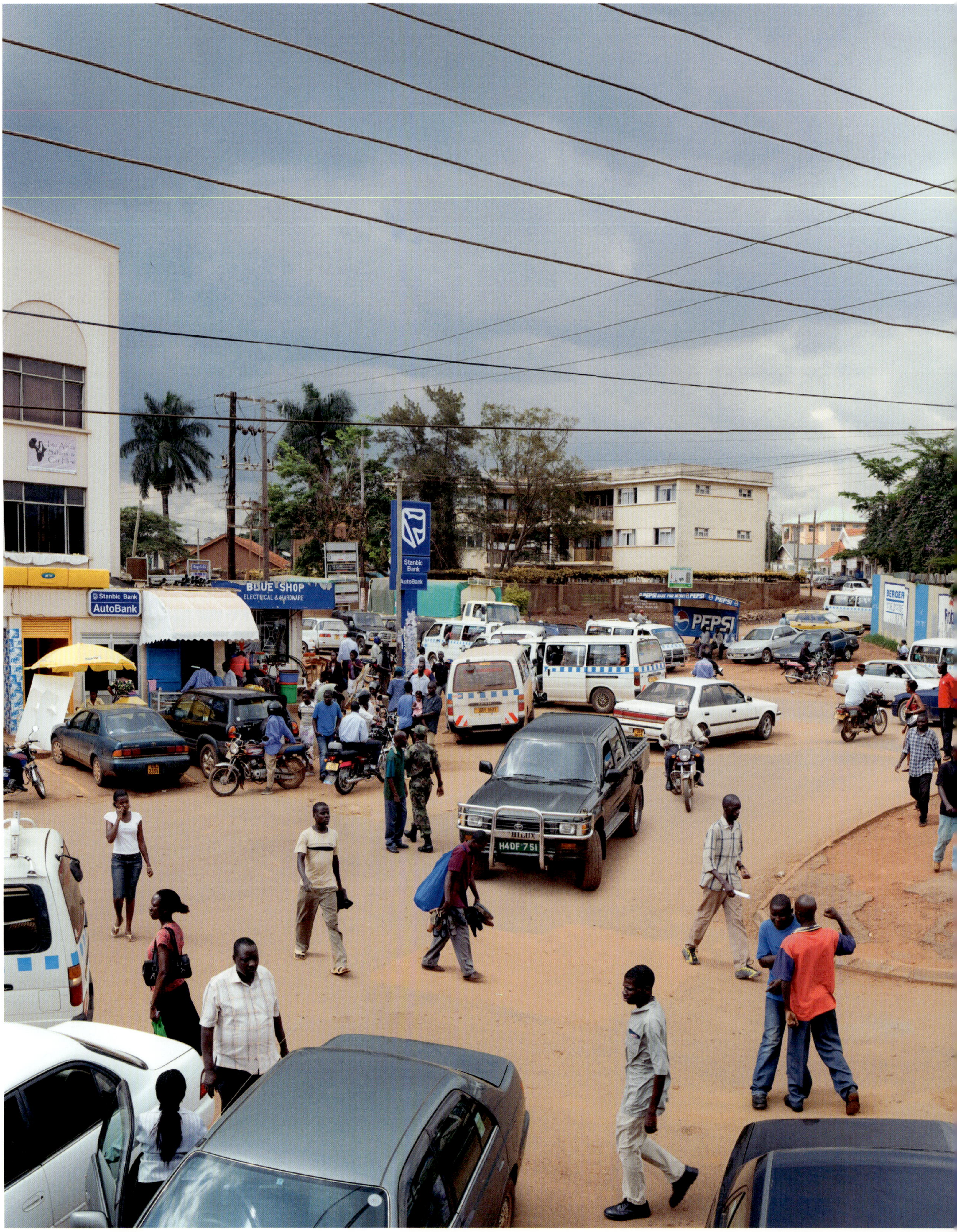
Stanbic Bank
AutoBank
BLUE SHOP
ELECTRICAL & HARDWARE
Stanbic Bank
AutoBank
PEPSI
BERGER
HILUX
H4DF 751

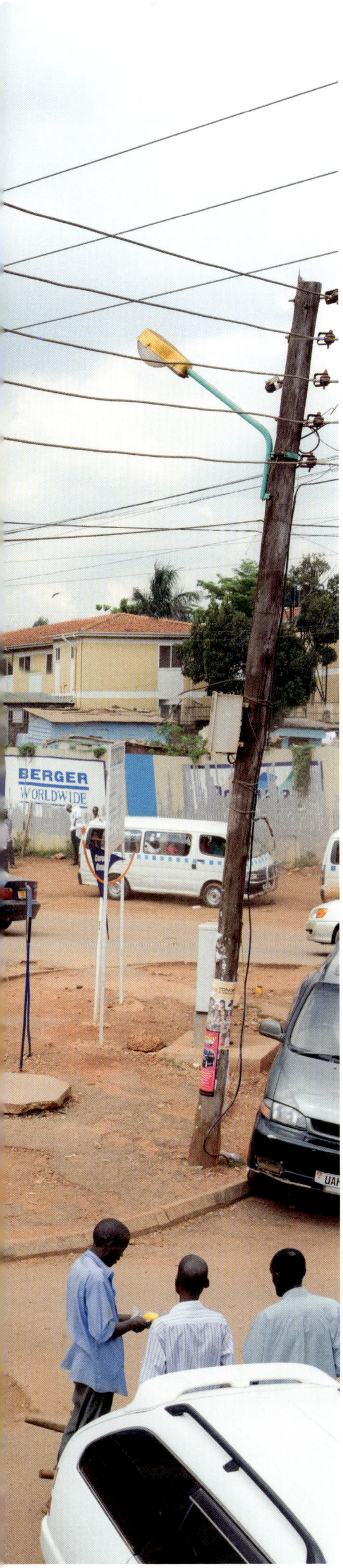

SPRING ROAD, BUGOLOGI
KAMPALA

RUA DU PATRICIO, CAZENGA
LUANDA

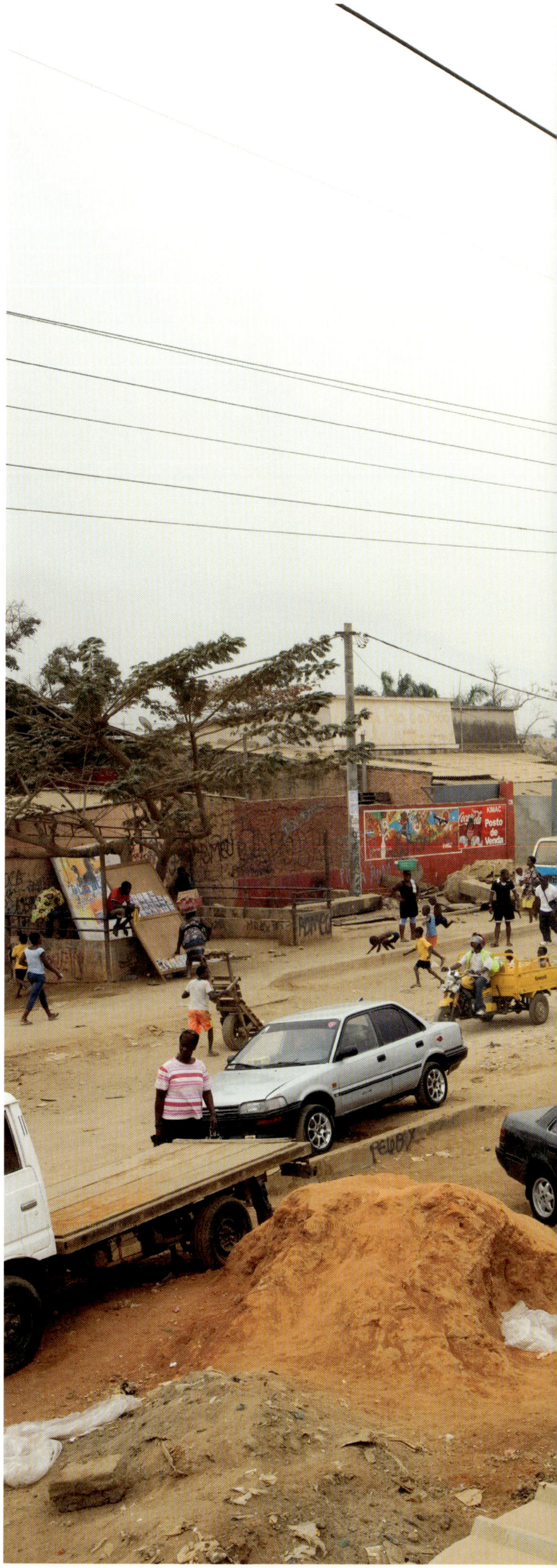

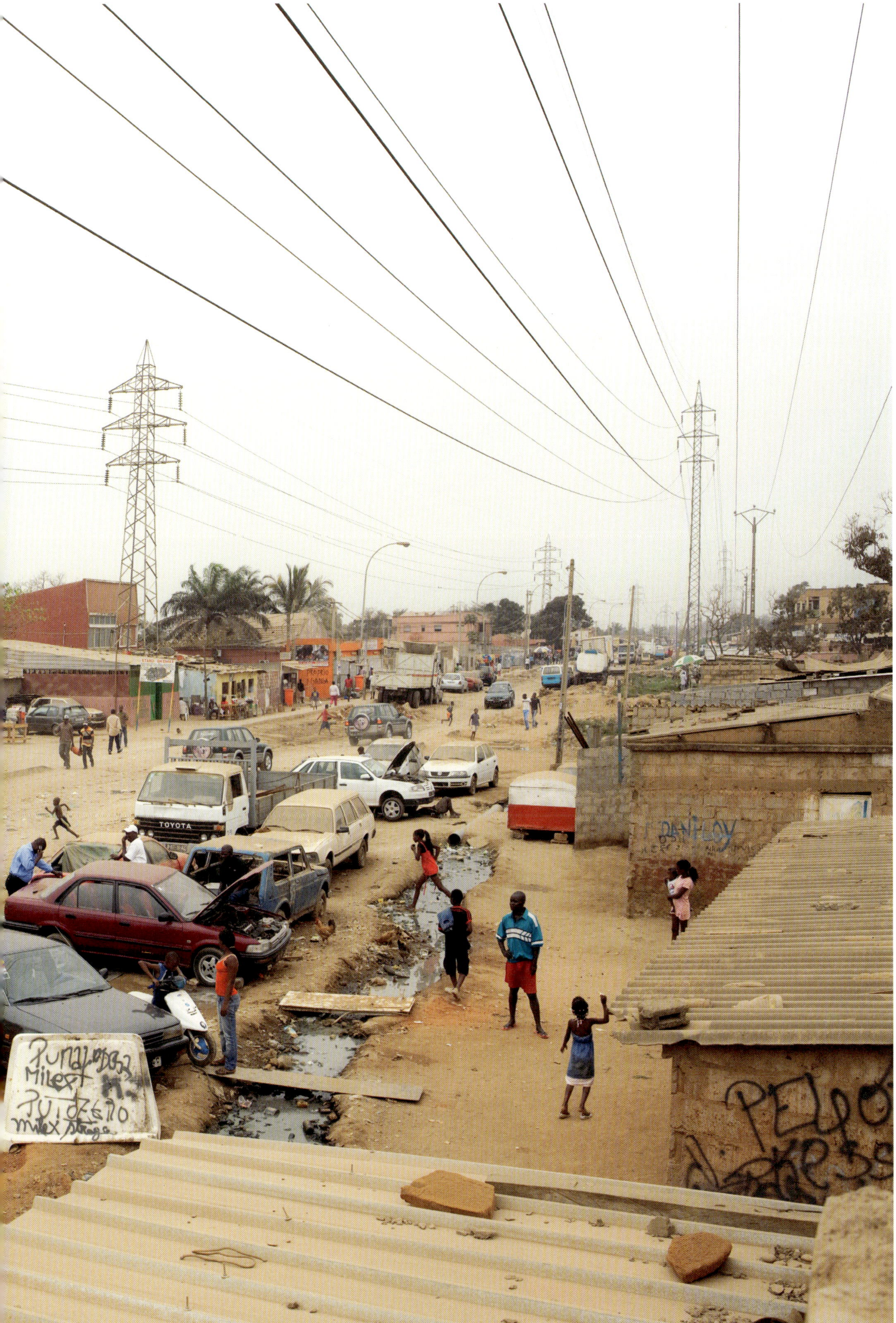
TOYOTA

BWAISE (CRANE)
KAMPALA

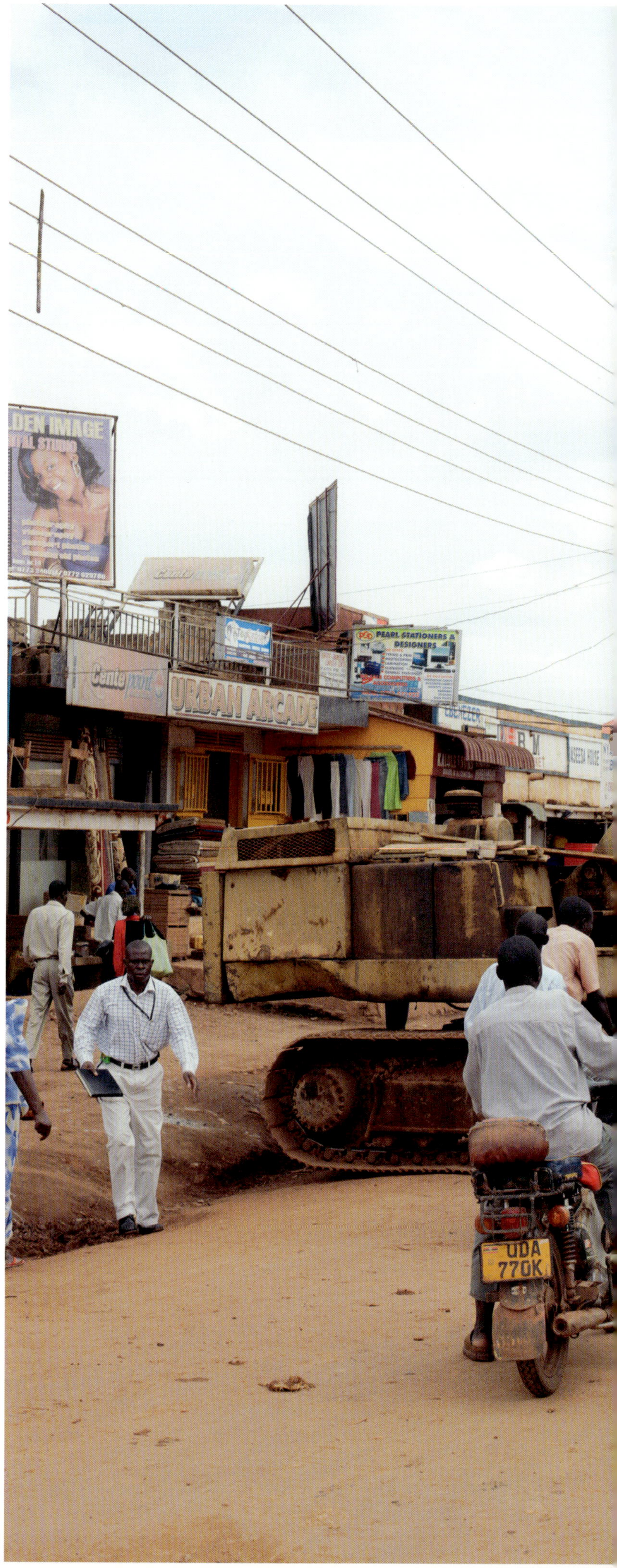

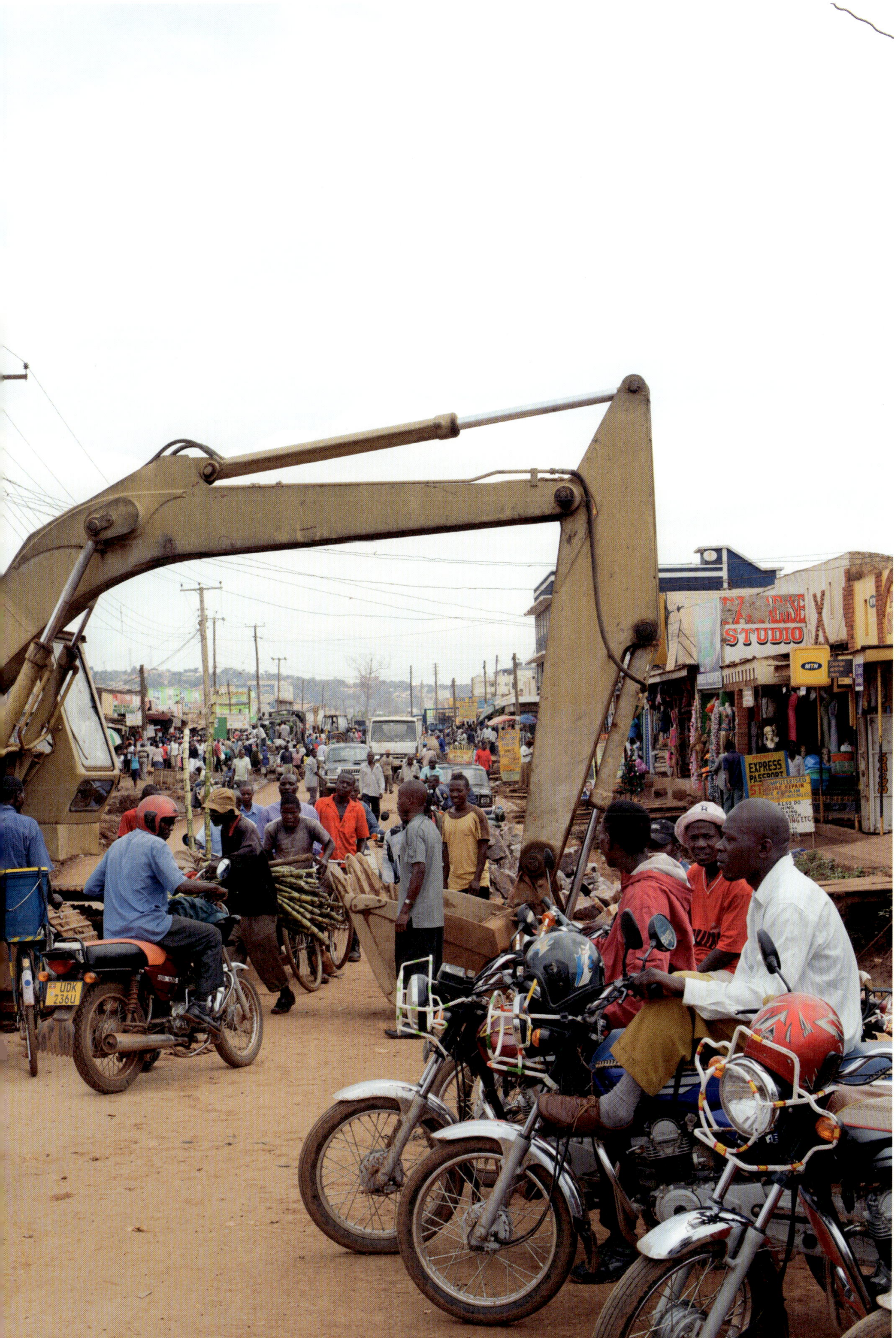
STUDIO
MTN
EXPRESS
UDK
236U

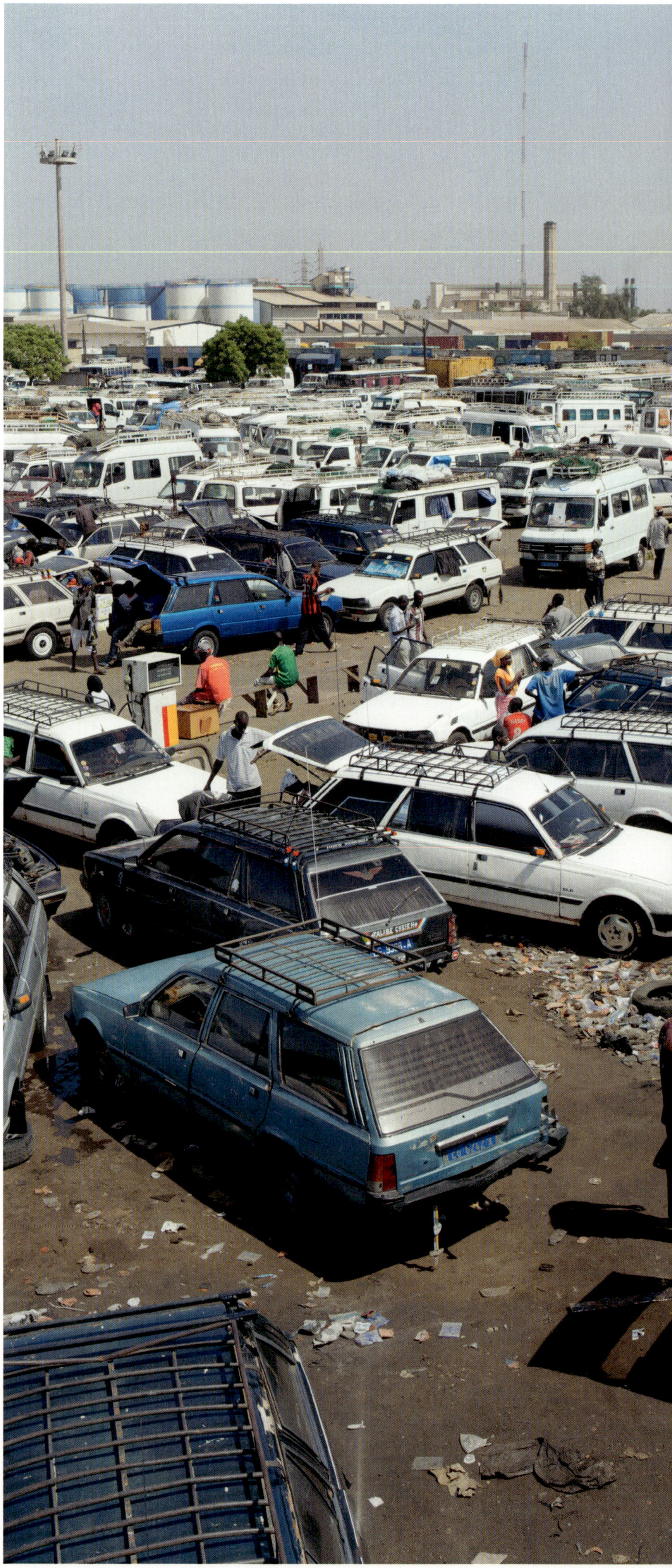

GARE ROUTIÈRE (POMPIERS)
DAKAR

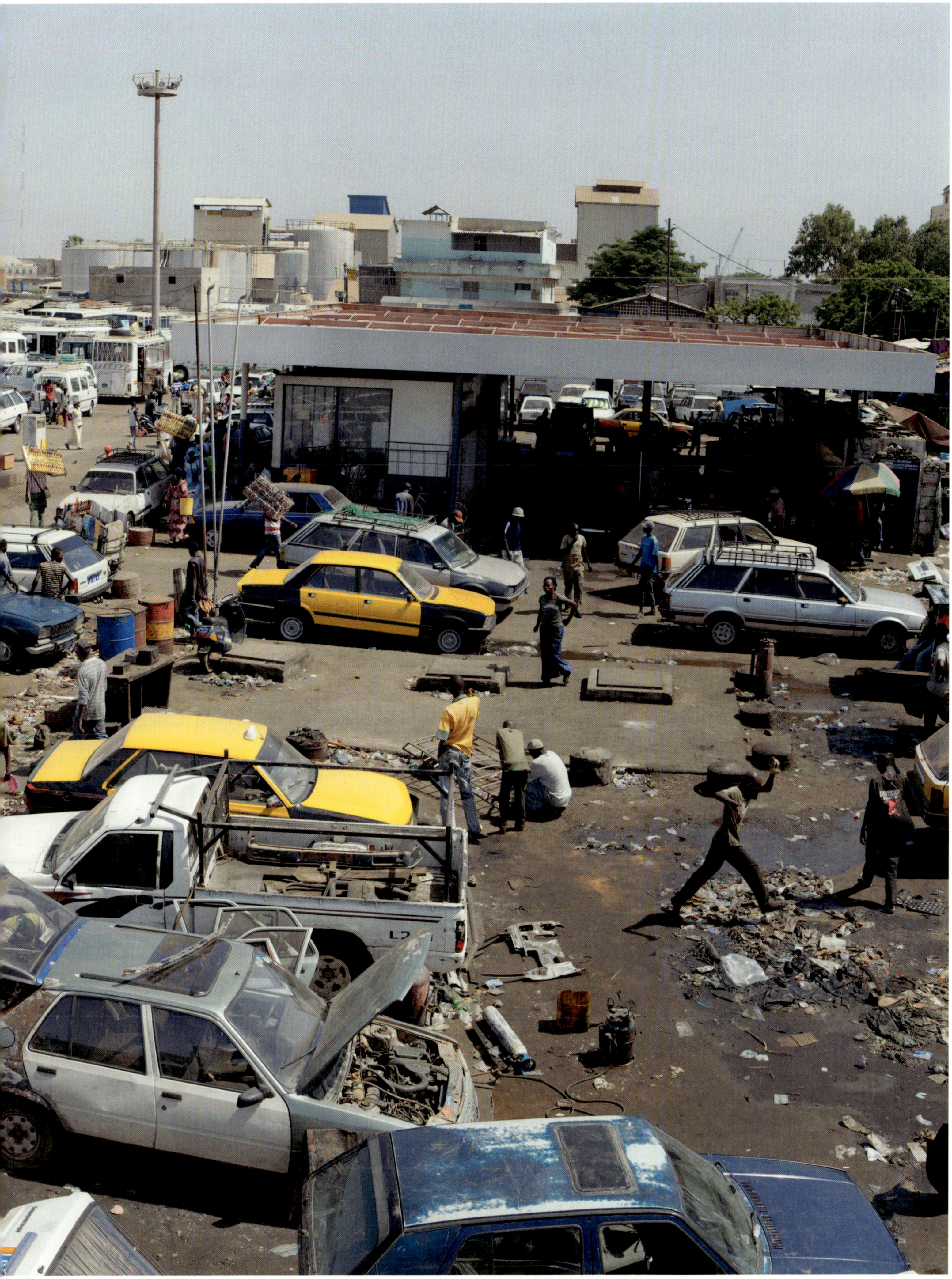

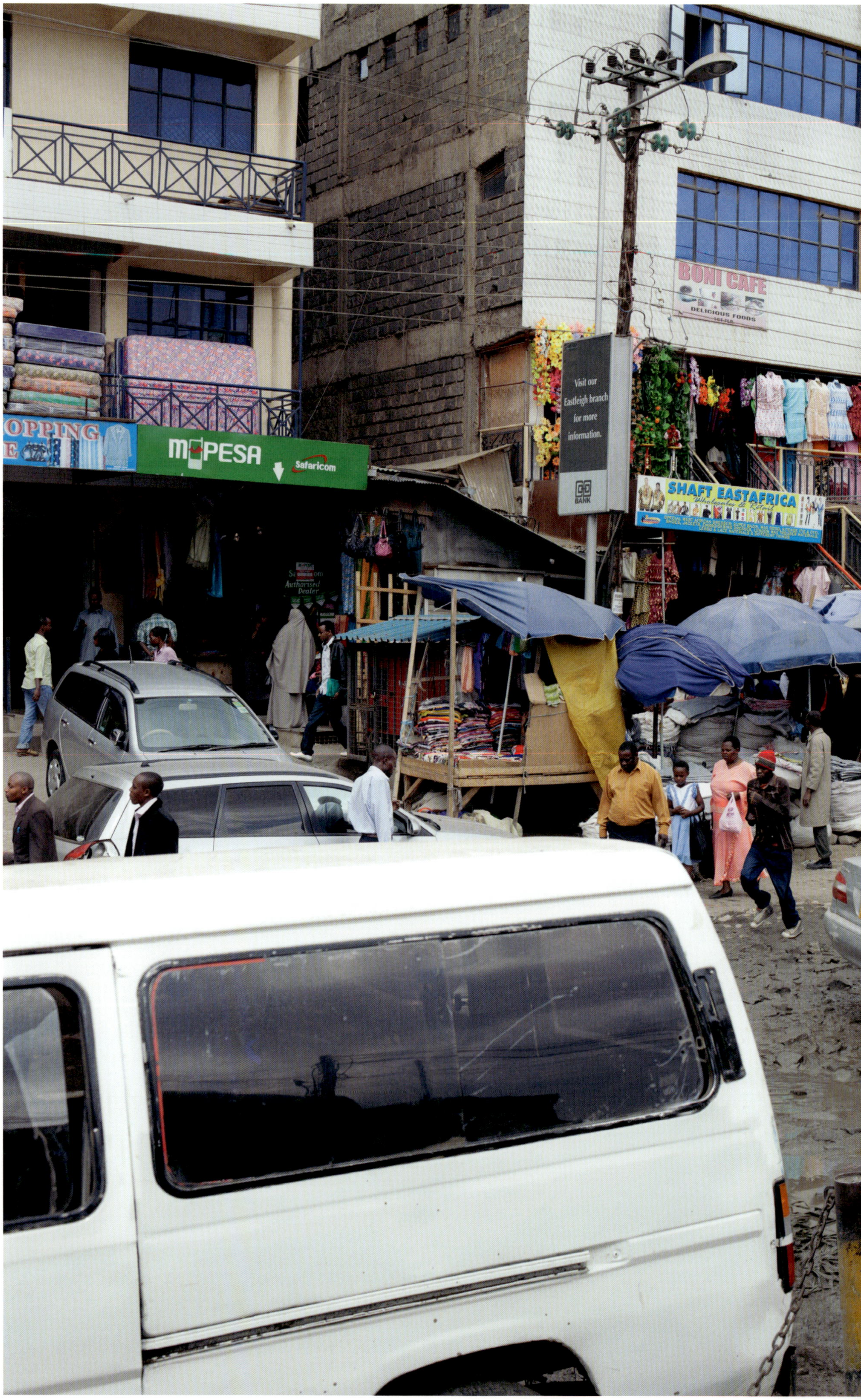

1ST AVENUE, EASTLEIGH
NAIROBI

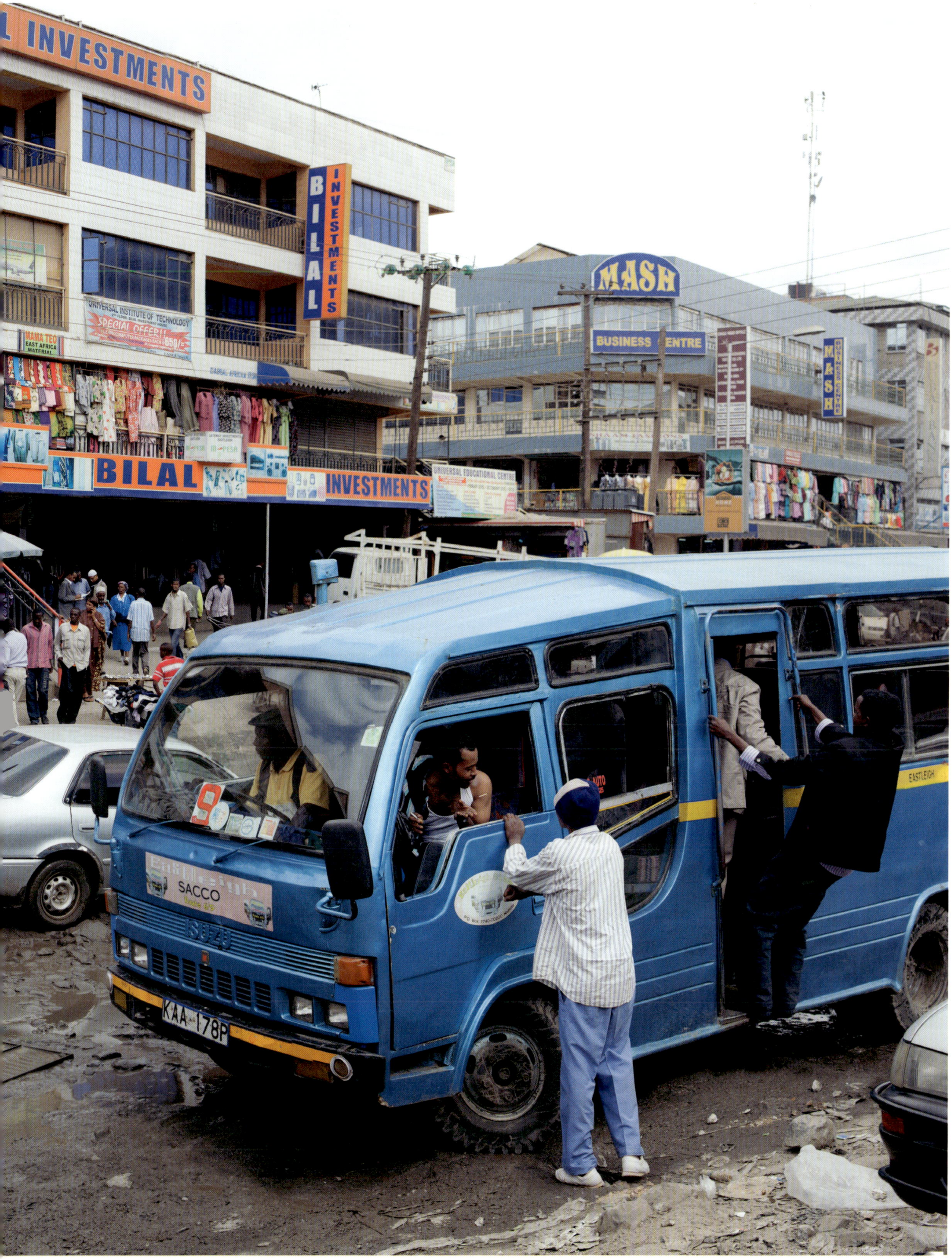

INVESTMENTS
BILAL INVESTMENTS
UNIVERSAL INSTITUTE OF TECHNOLOGY
SPECIAL OFFER!!
MAMA TEO
EAST AFRICA MATERIAL
BILAL
INVESTMENTS
UNIVERSAL EDUCATIONAL CENTRE
MASH
BUSINESS CENTRE
MASH
SACCO
ISUZU
KAA 178P
EASTLEIGH

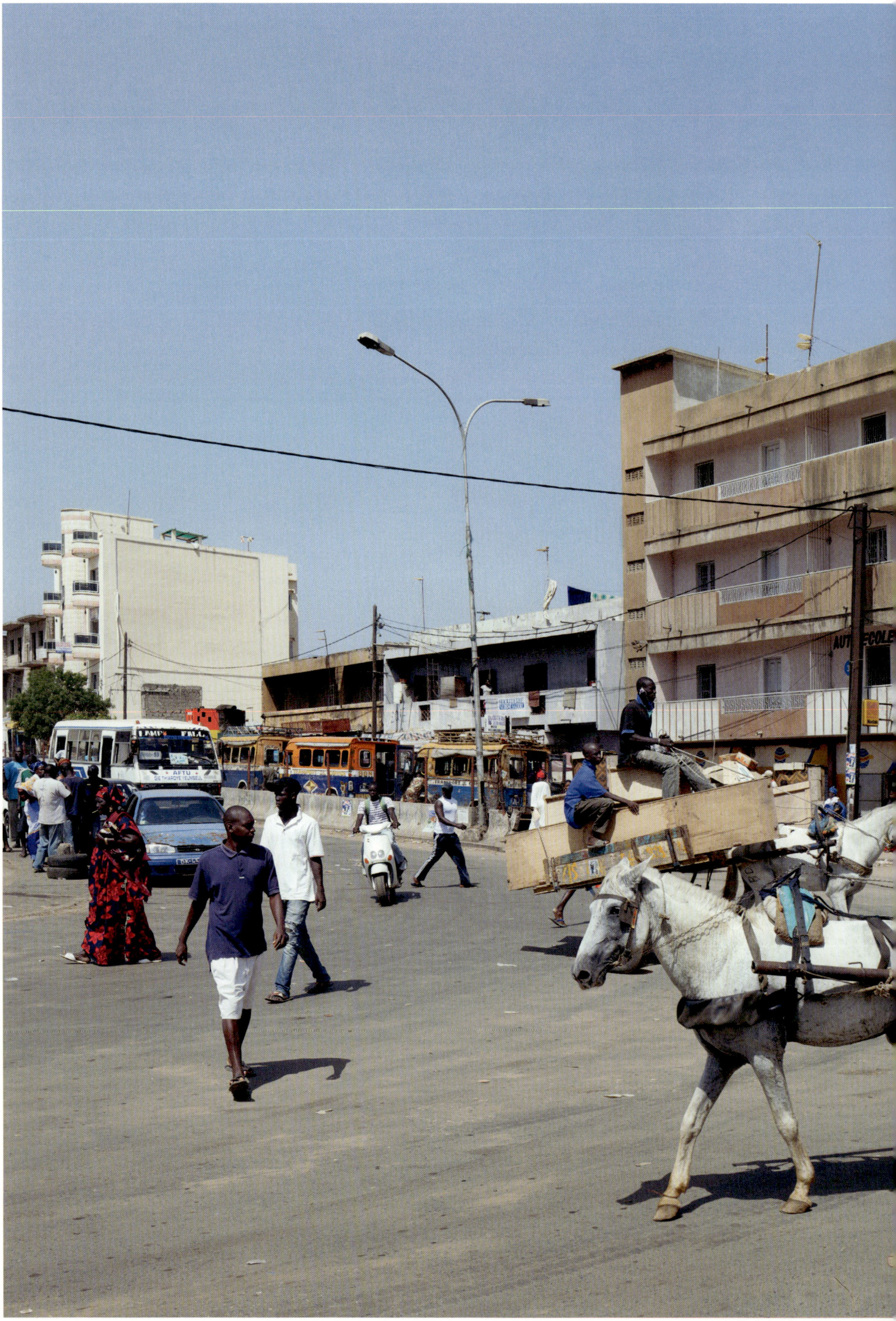

CROISEMENT TALLY BOUMAK
DAKAR

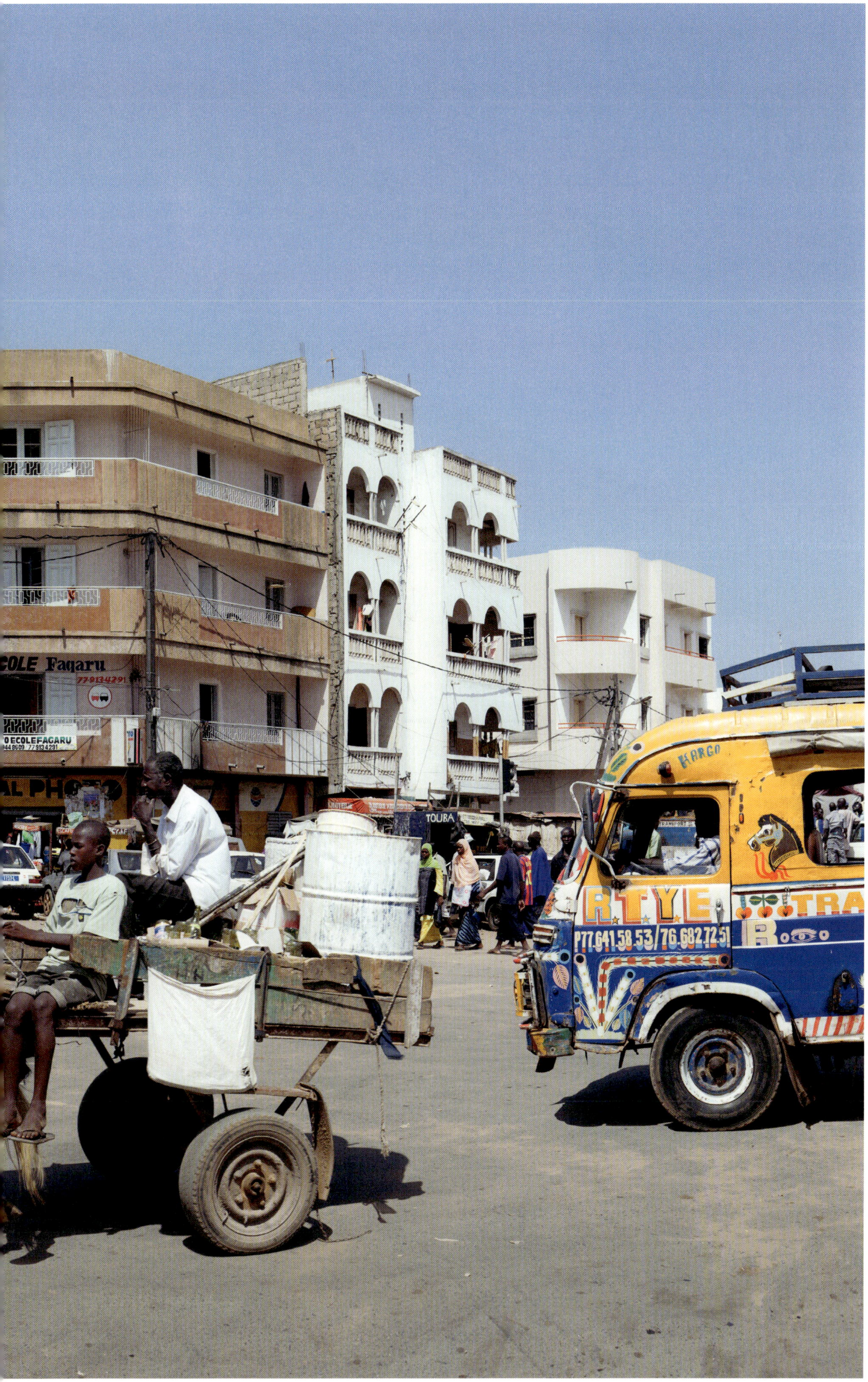
COLE Fagaru
TOUBA
KARGO
RTYE
TRA

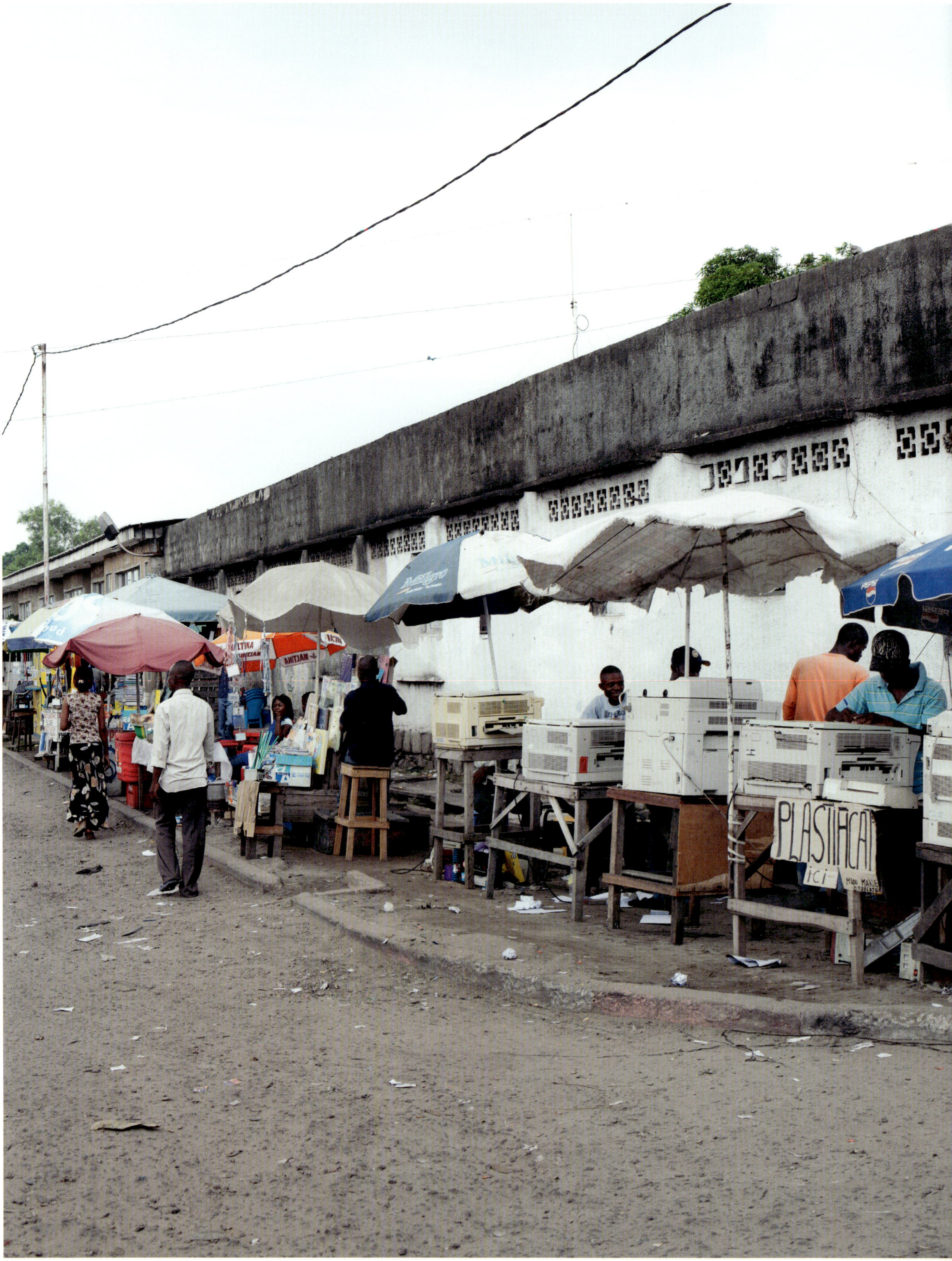
PLASTIFICAT
ici

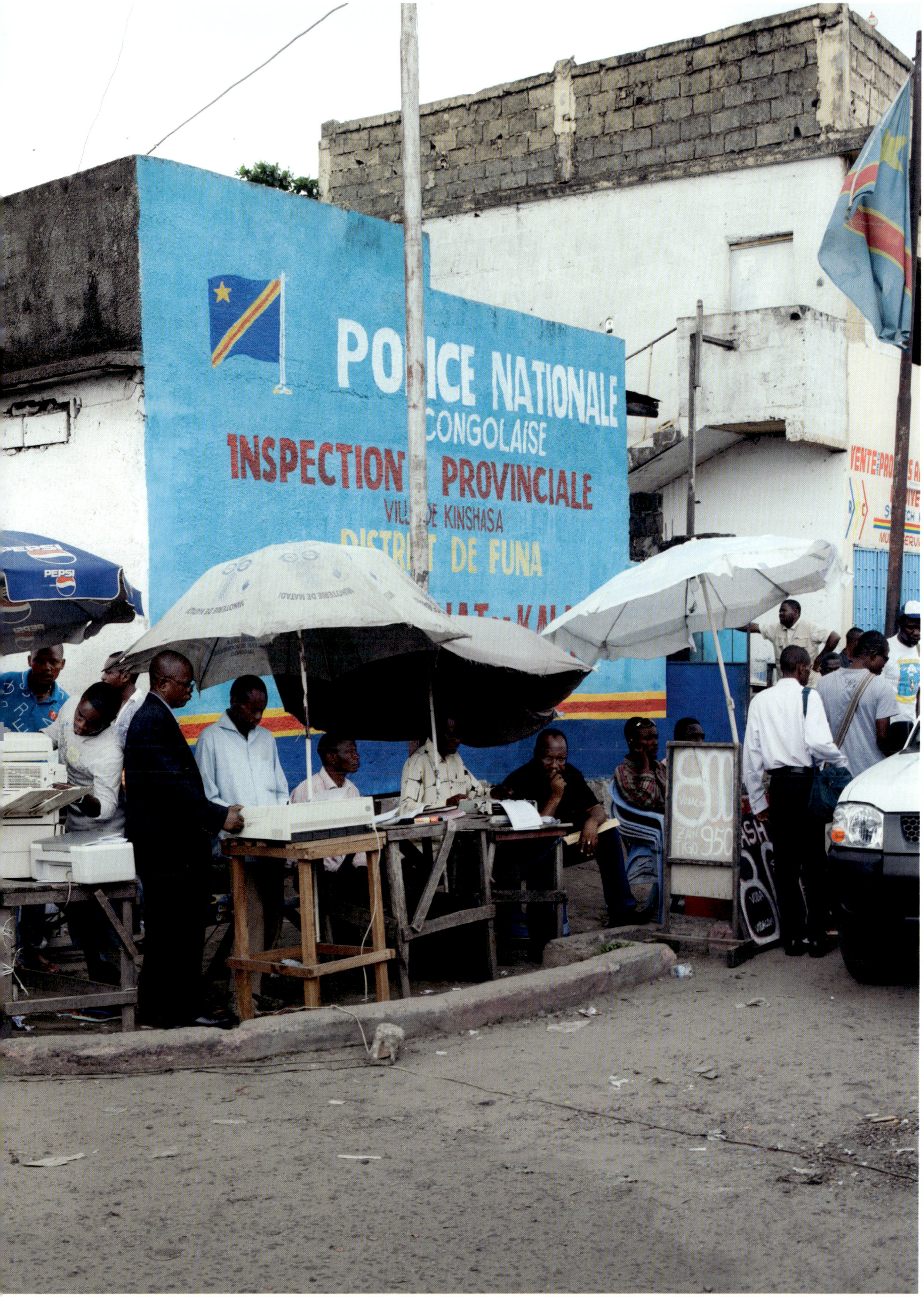

AVENUE BOKOKO
KINSHASA

U.P.N. UNIVERSITÉ
KINSHASA

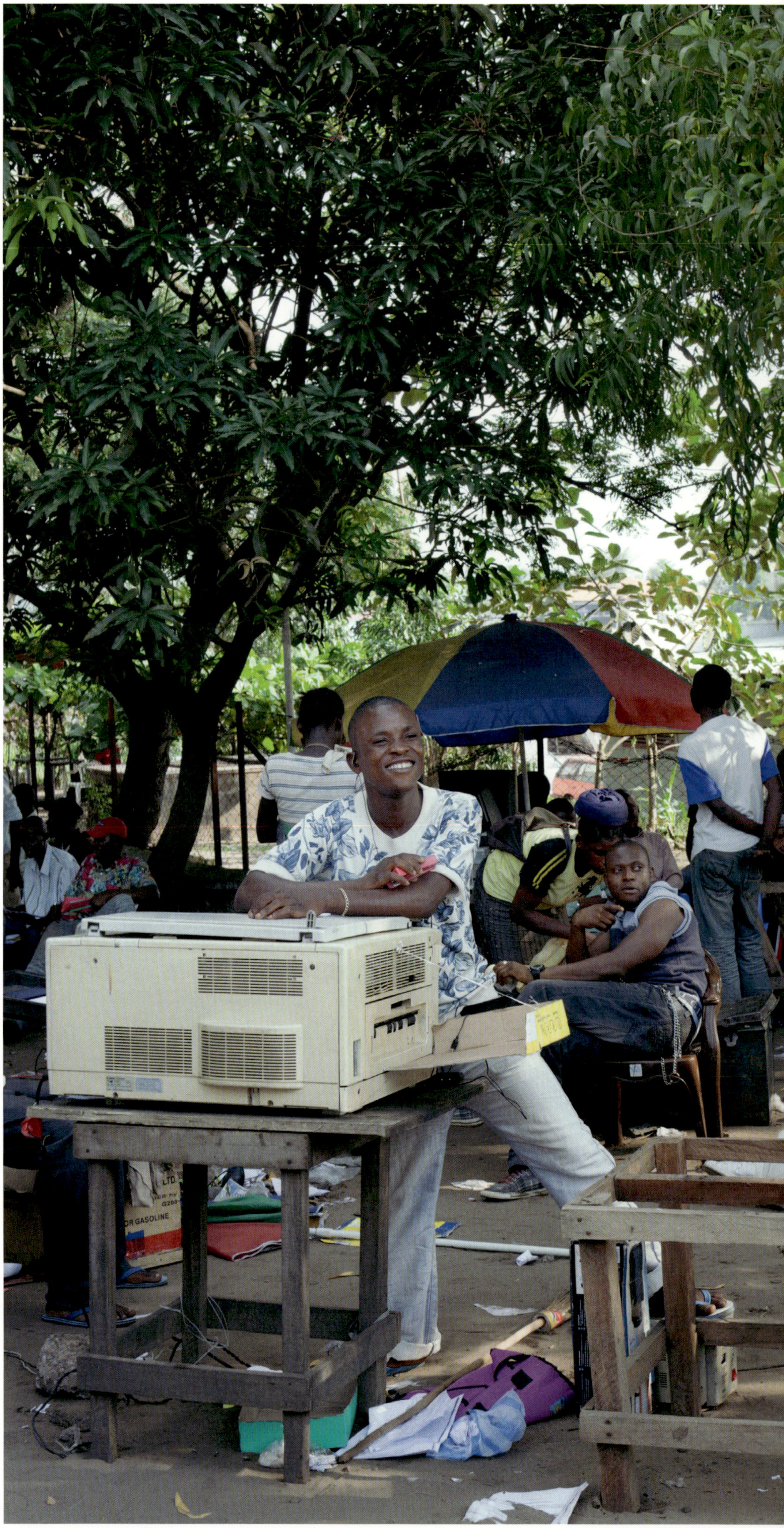

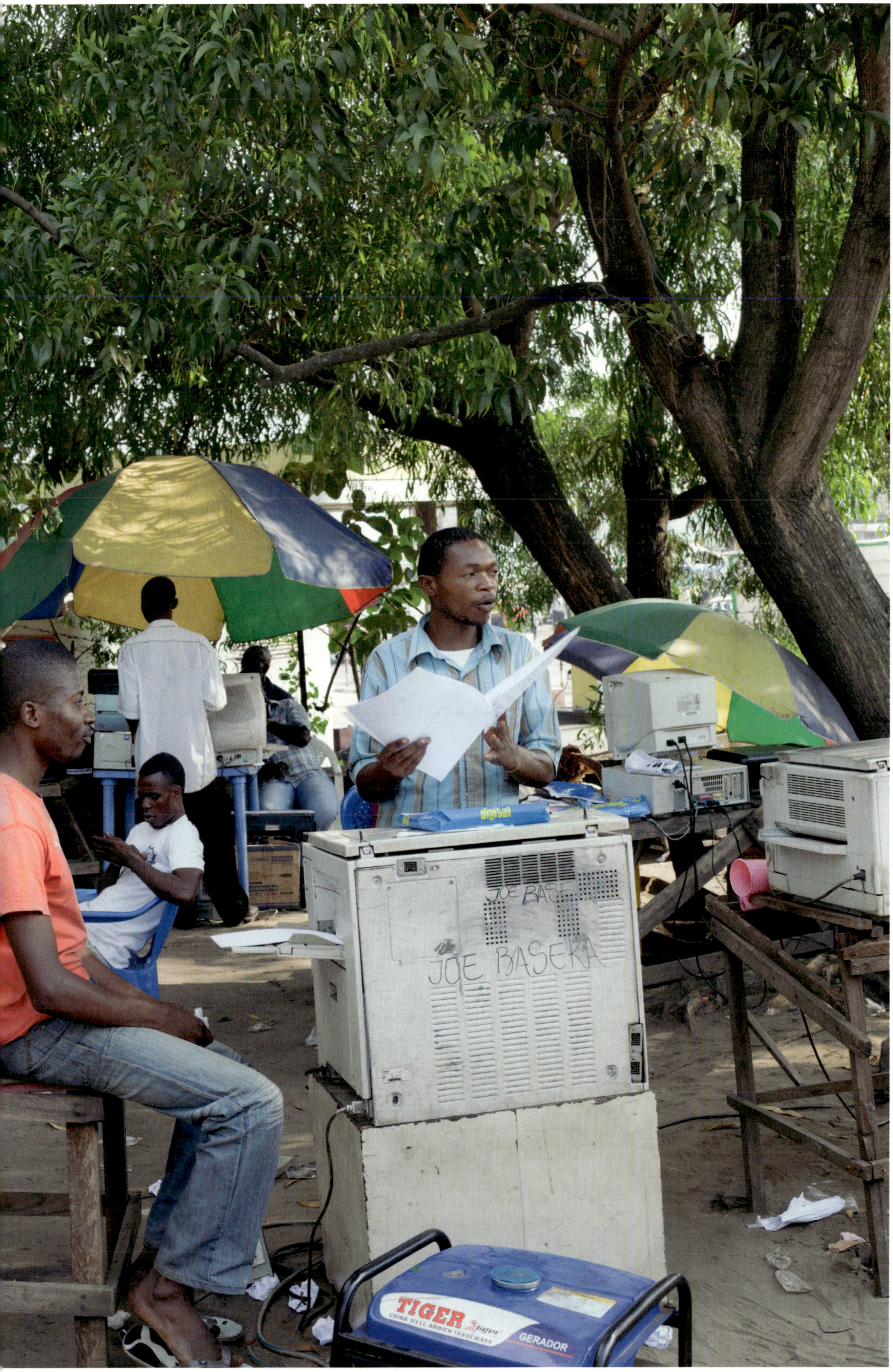
digital
JOE BASEKA
TIGER
CHINA WELL KNOWN TRADE MARK
GERADOR

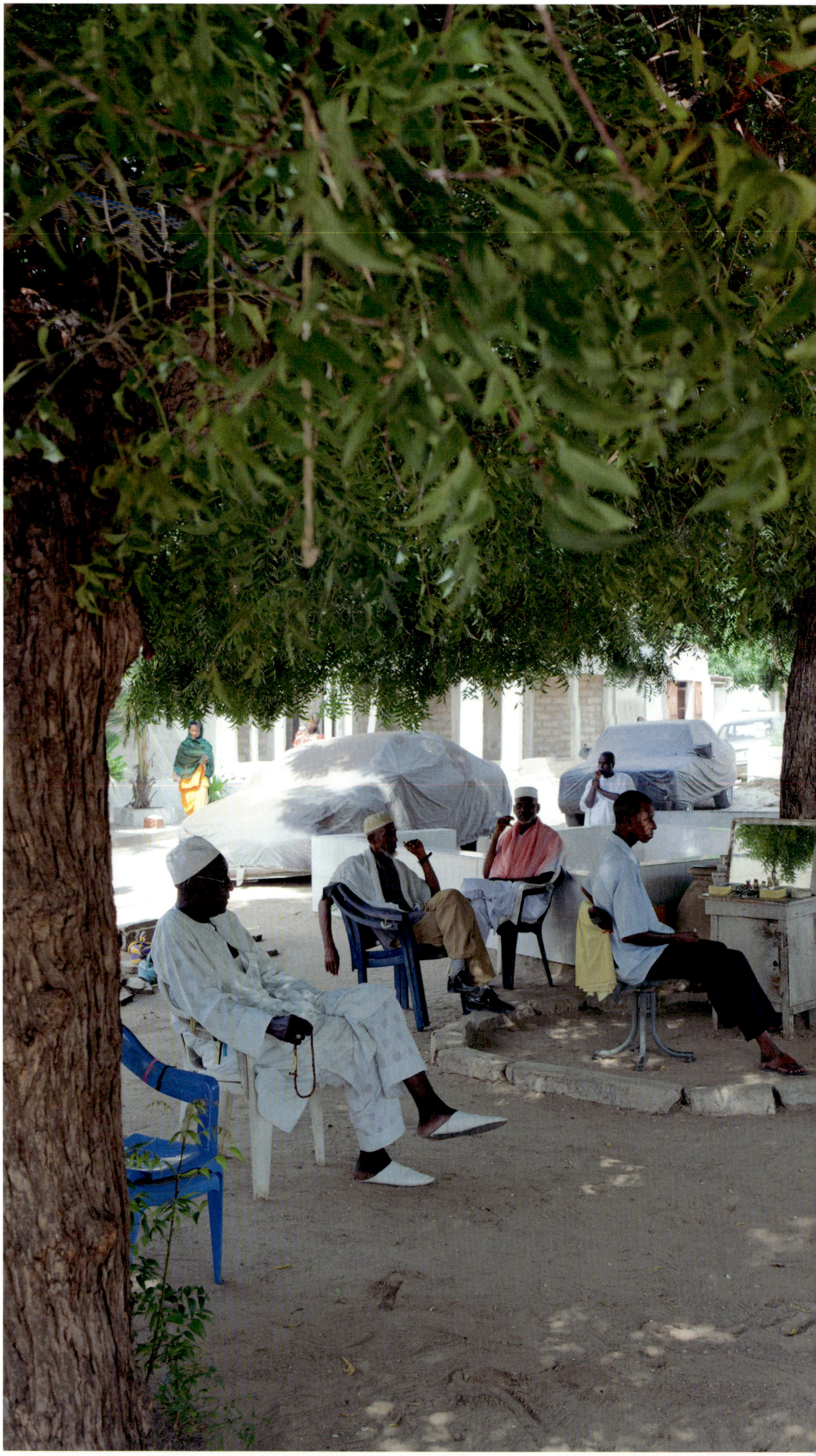

S.I.C.A.P. RUE DIX
DAKAR

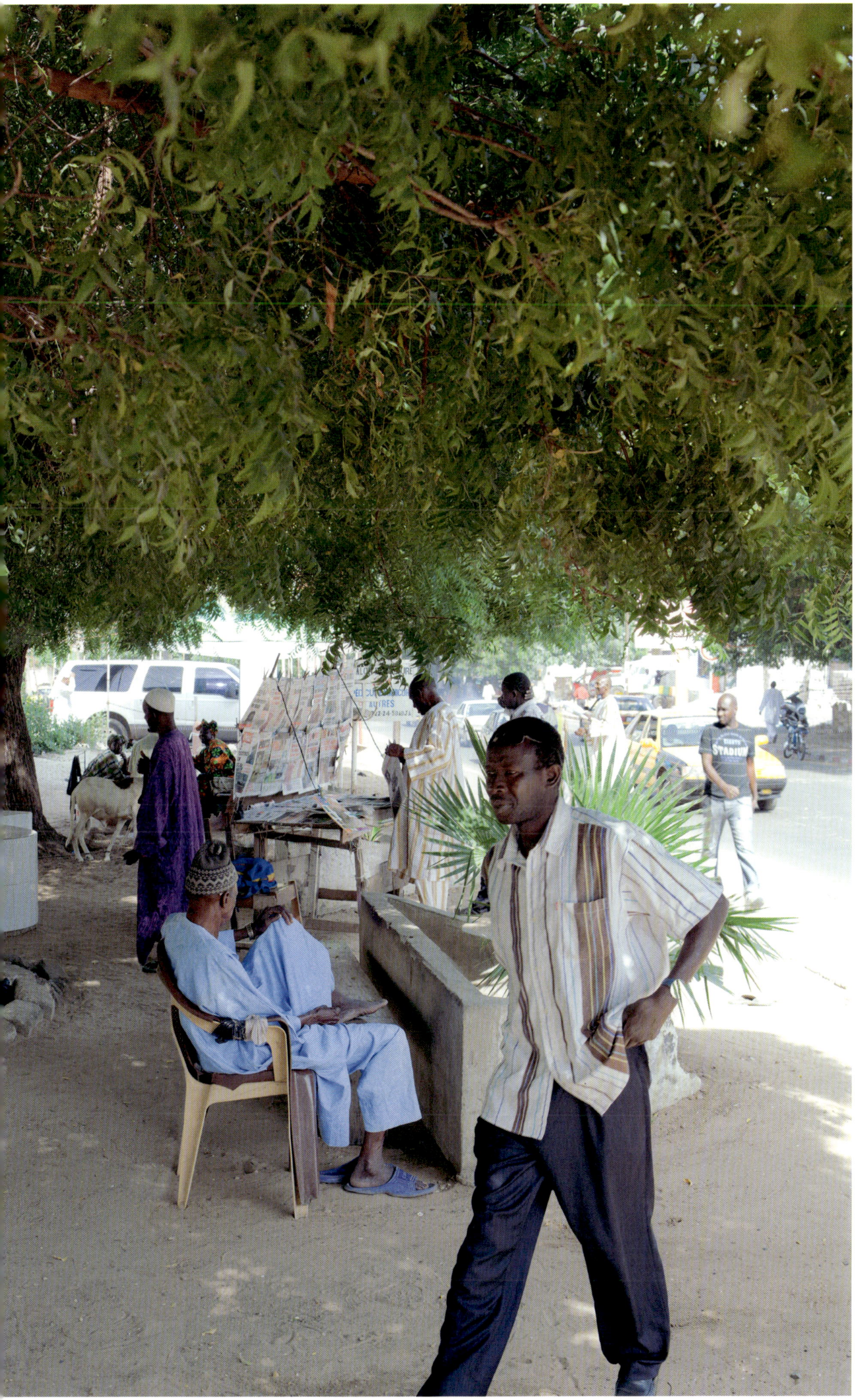
AUTRES
STADIUM

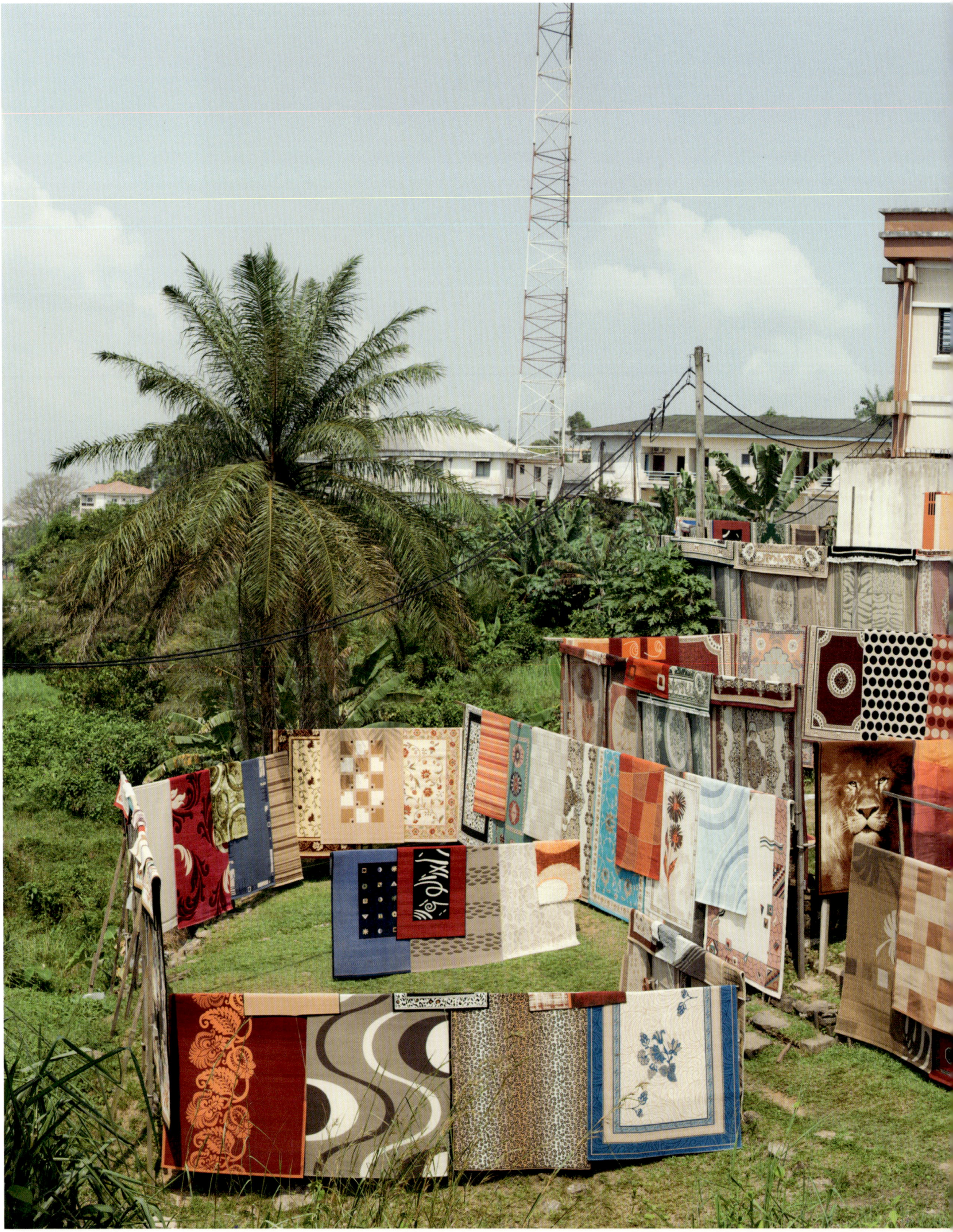

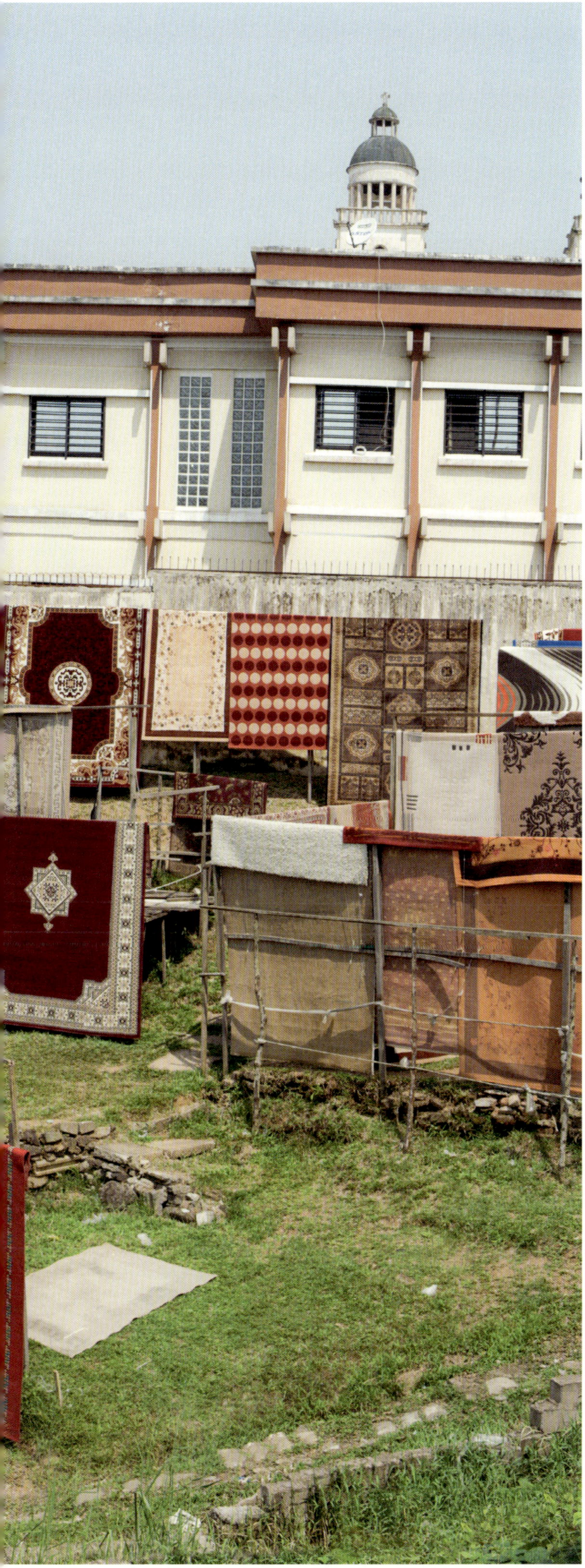

ROND-POINT JOSS
DOUALA

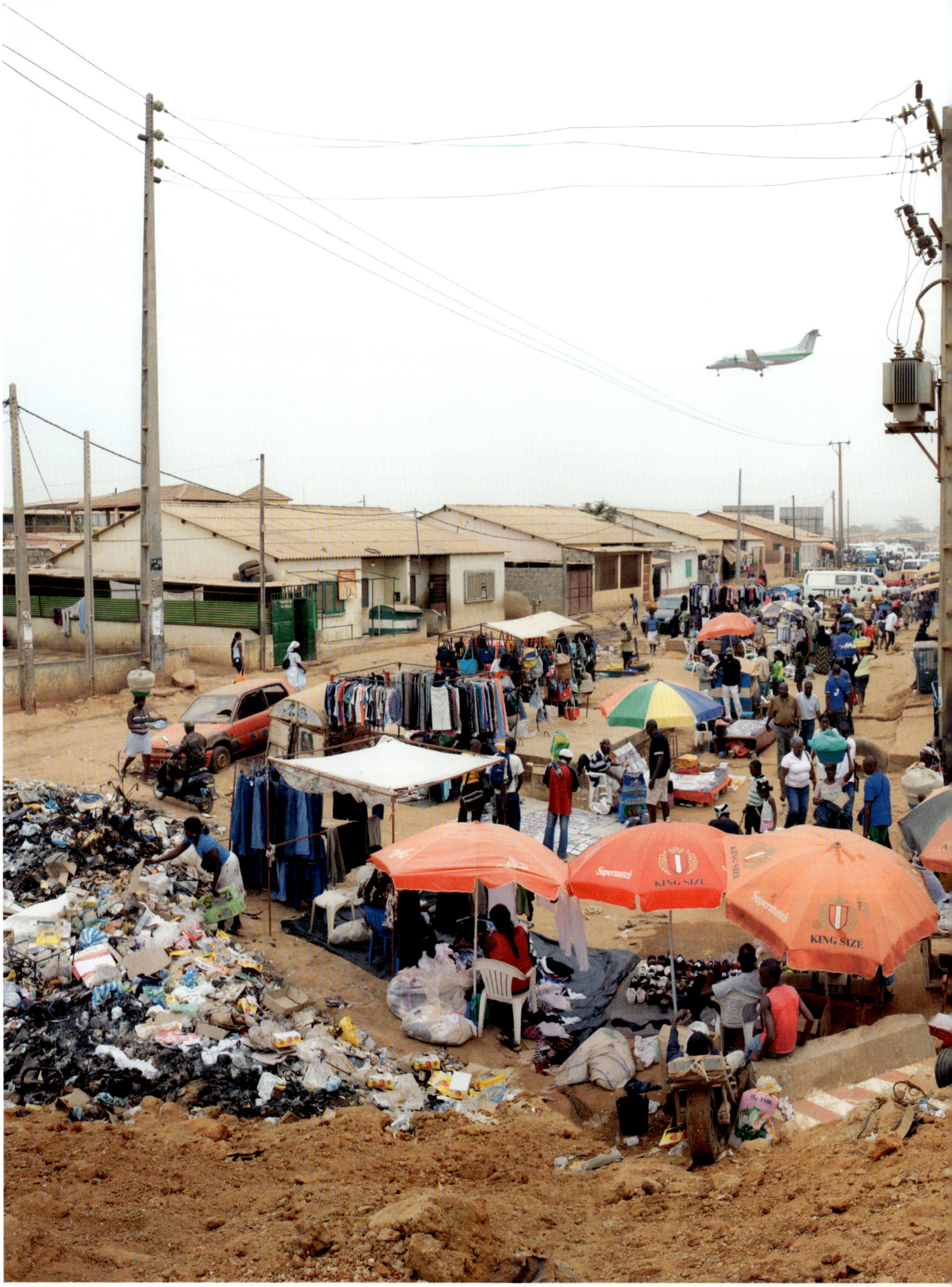
Supermatch
KING SIZE
Supermatch
KING SIZE

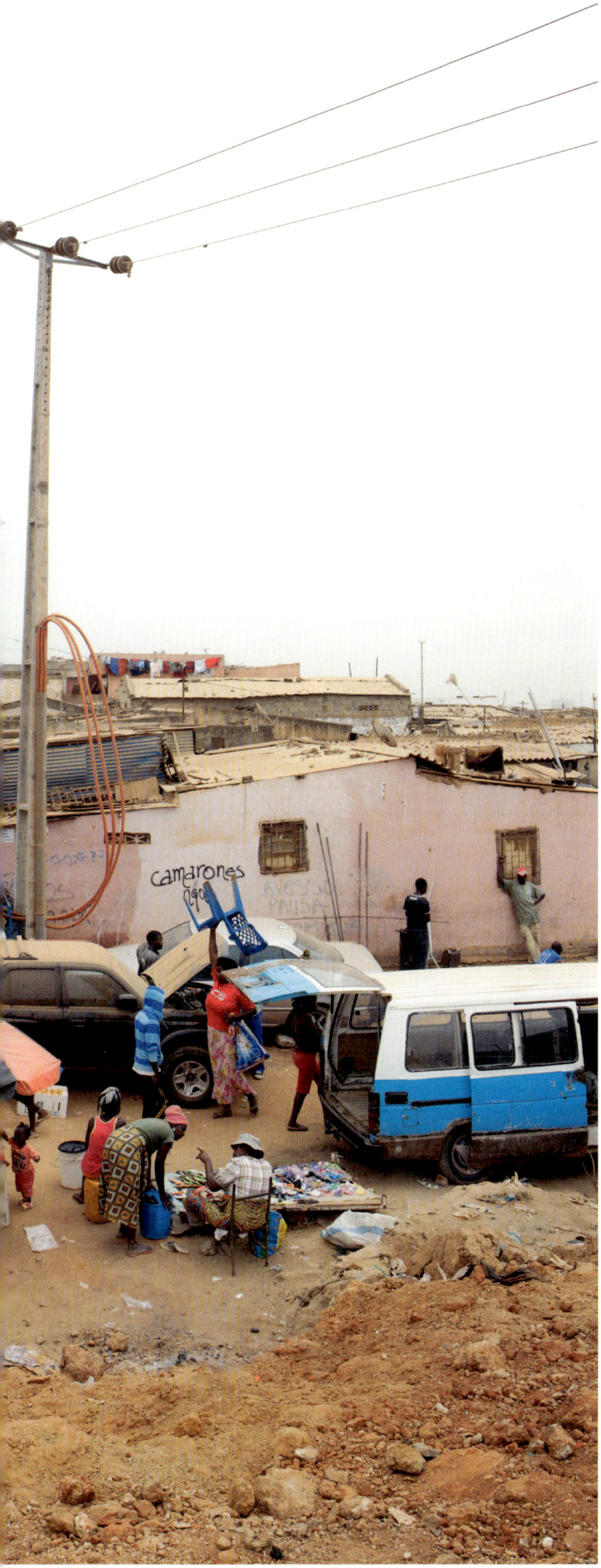

BAIRRO CASAQUELE
LUANDA

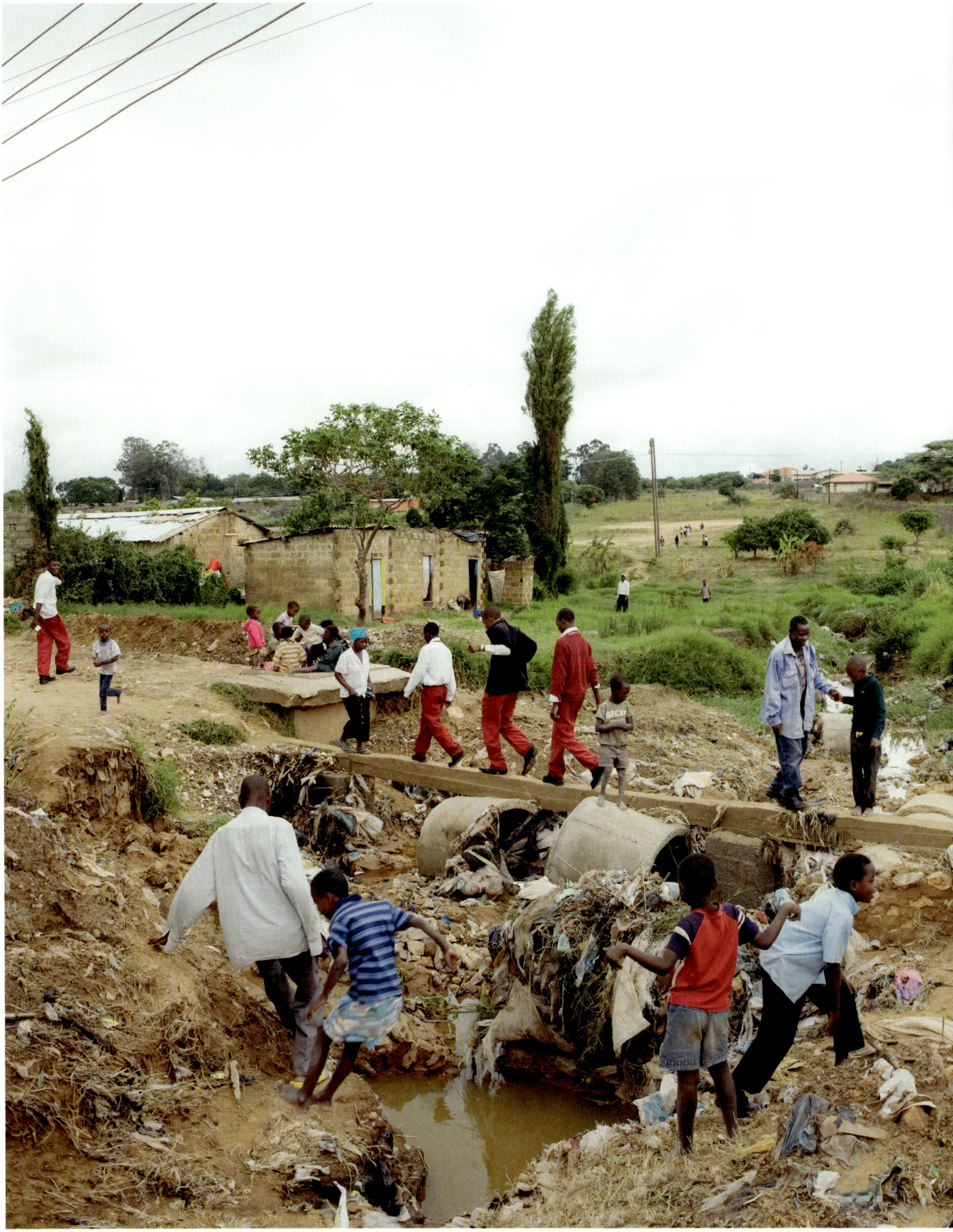

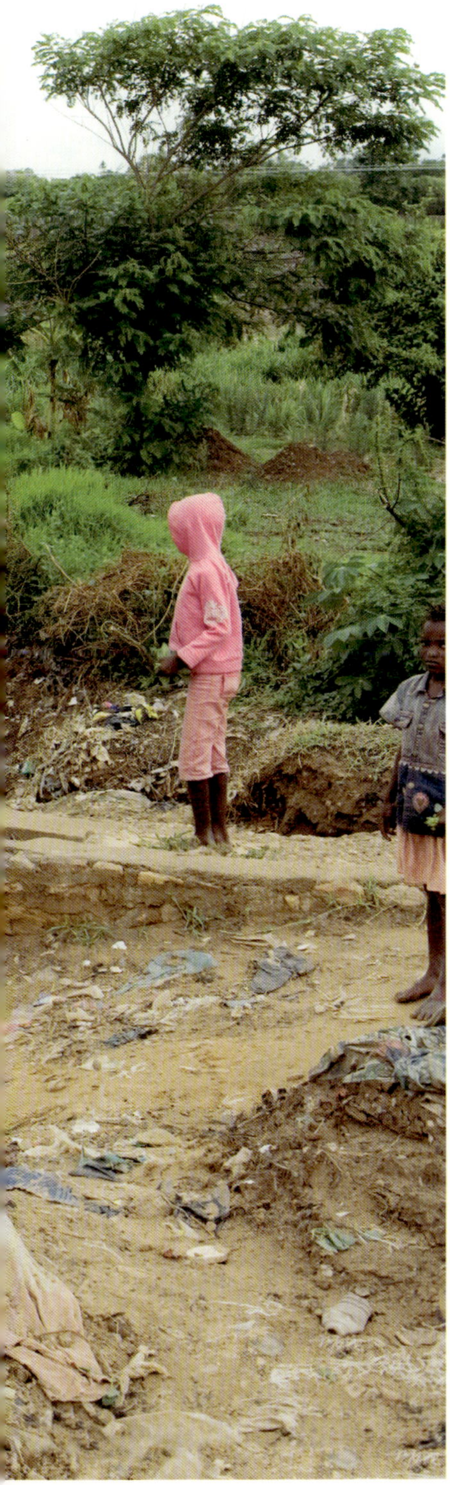

AISHA ROAD, N'GOMBE
LUSAKA

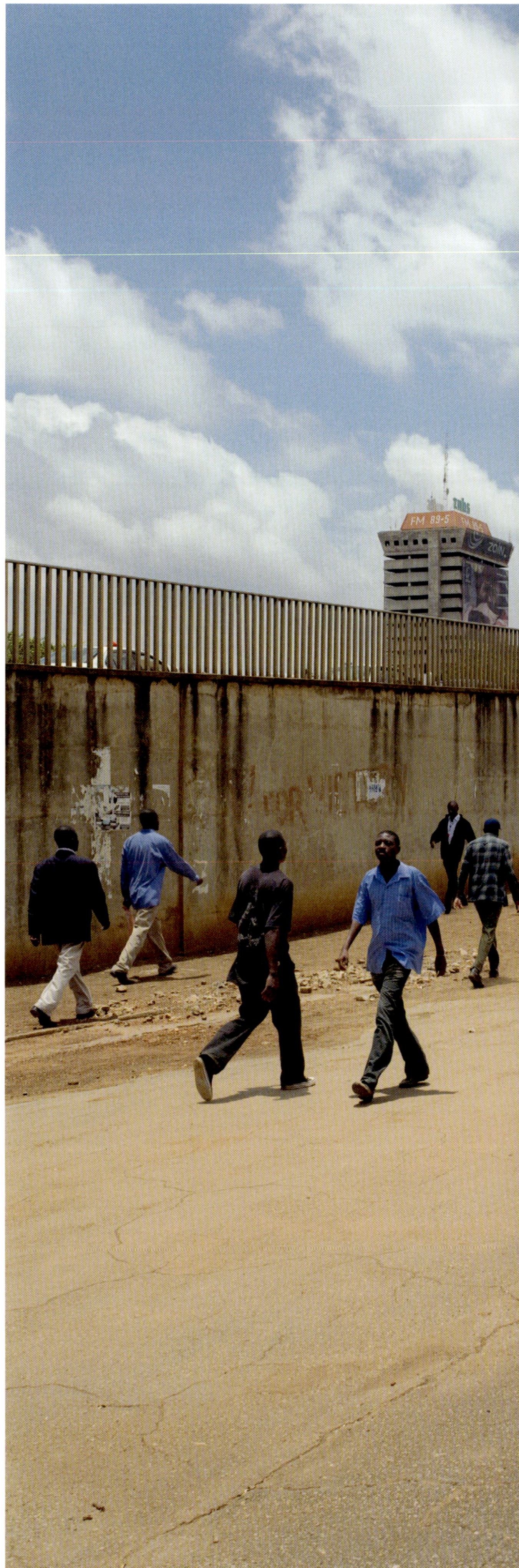

CHURCH ROAD (FLY-OVER)
LUSAKA

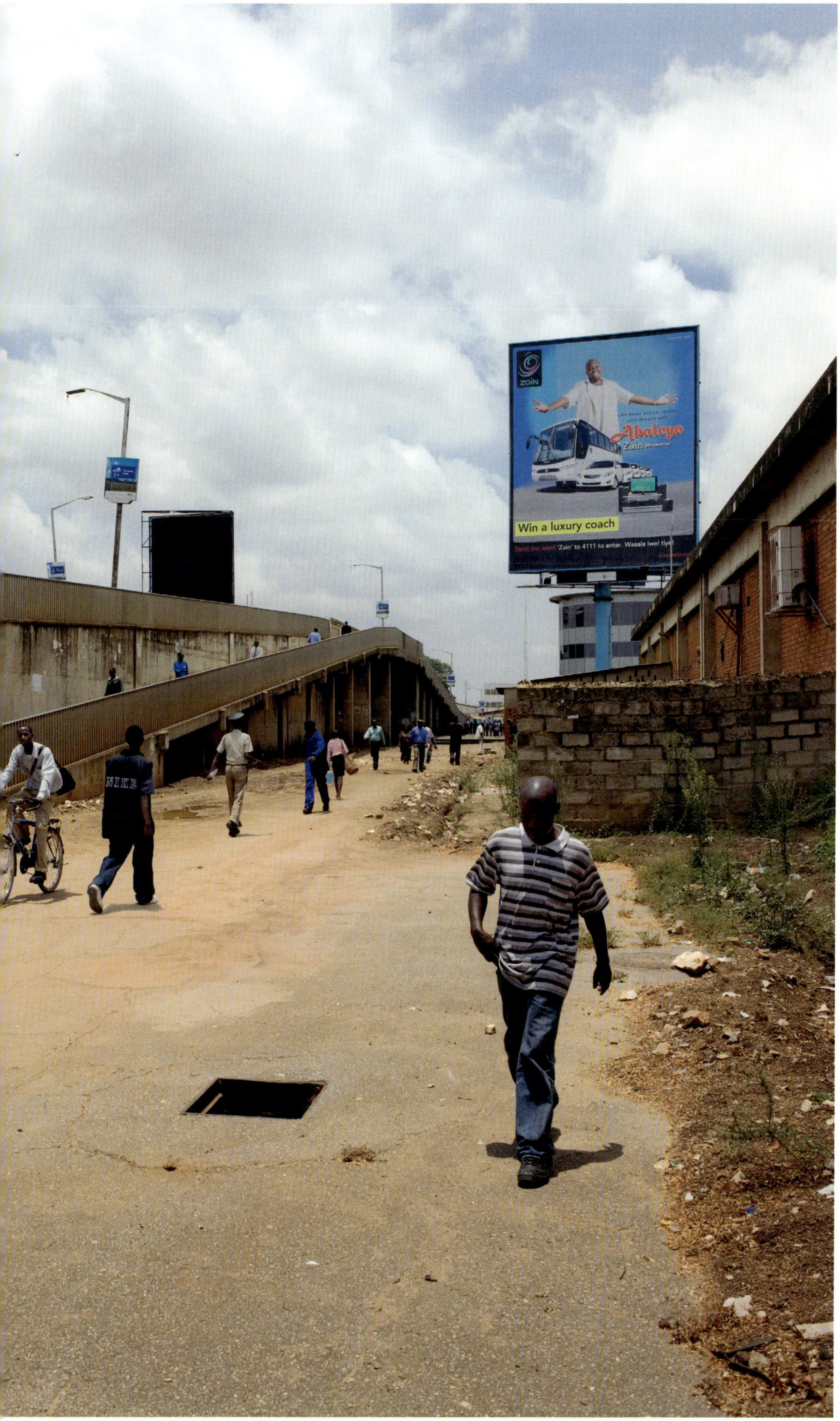
zain
Like never before, realise your dreams with
Abaleya
Zain Promotion
Win a luxury coach
Send the word 'Zain' to 4111 to enter. Wasala iwe! tiye!
NIKE

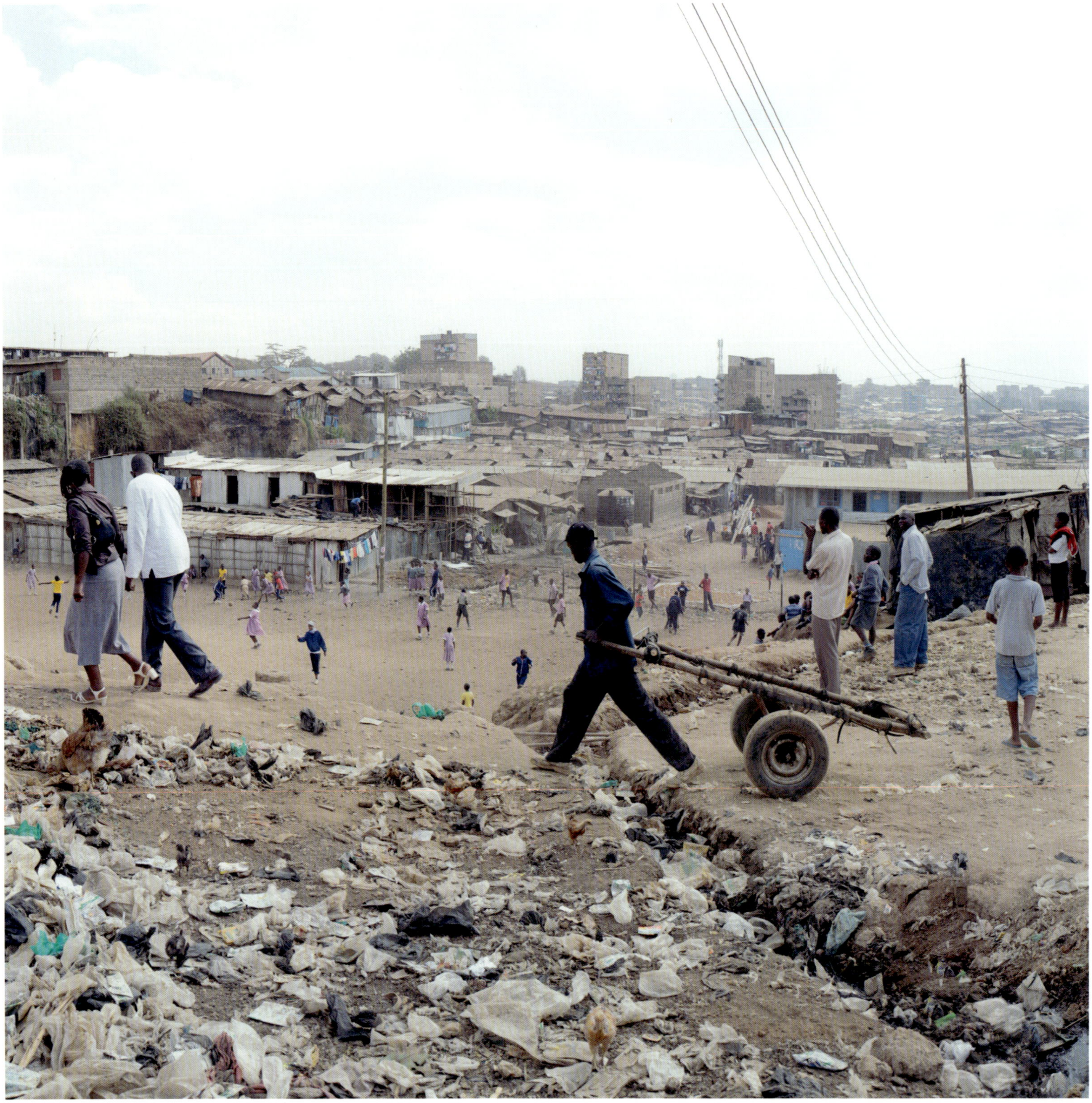

SOCCER FIELD, MATHARE
NAIROBI

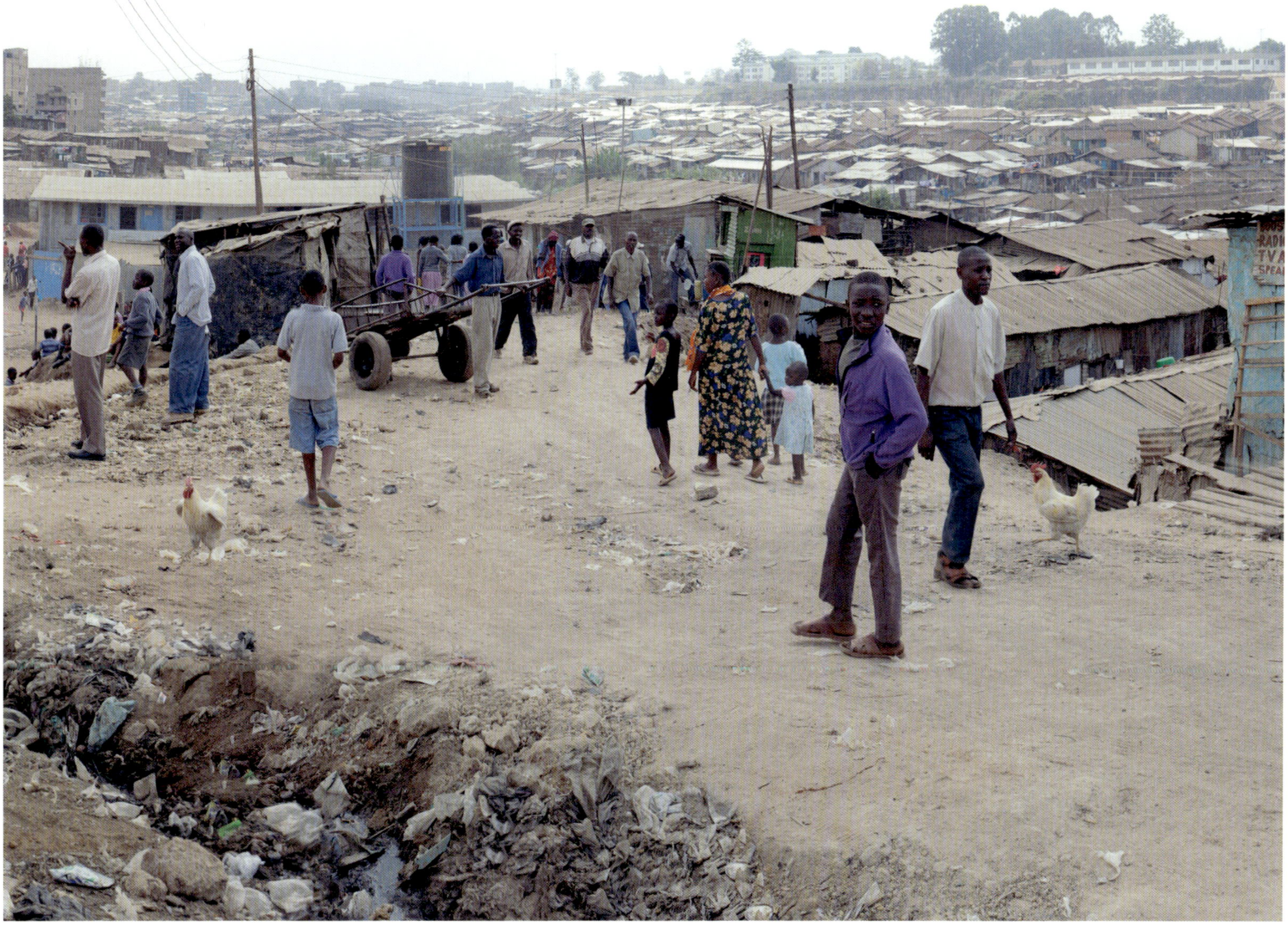

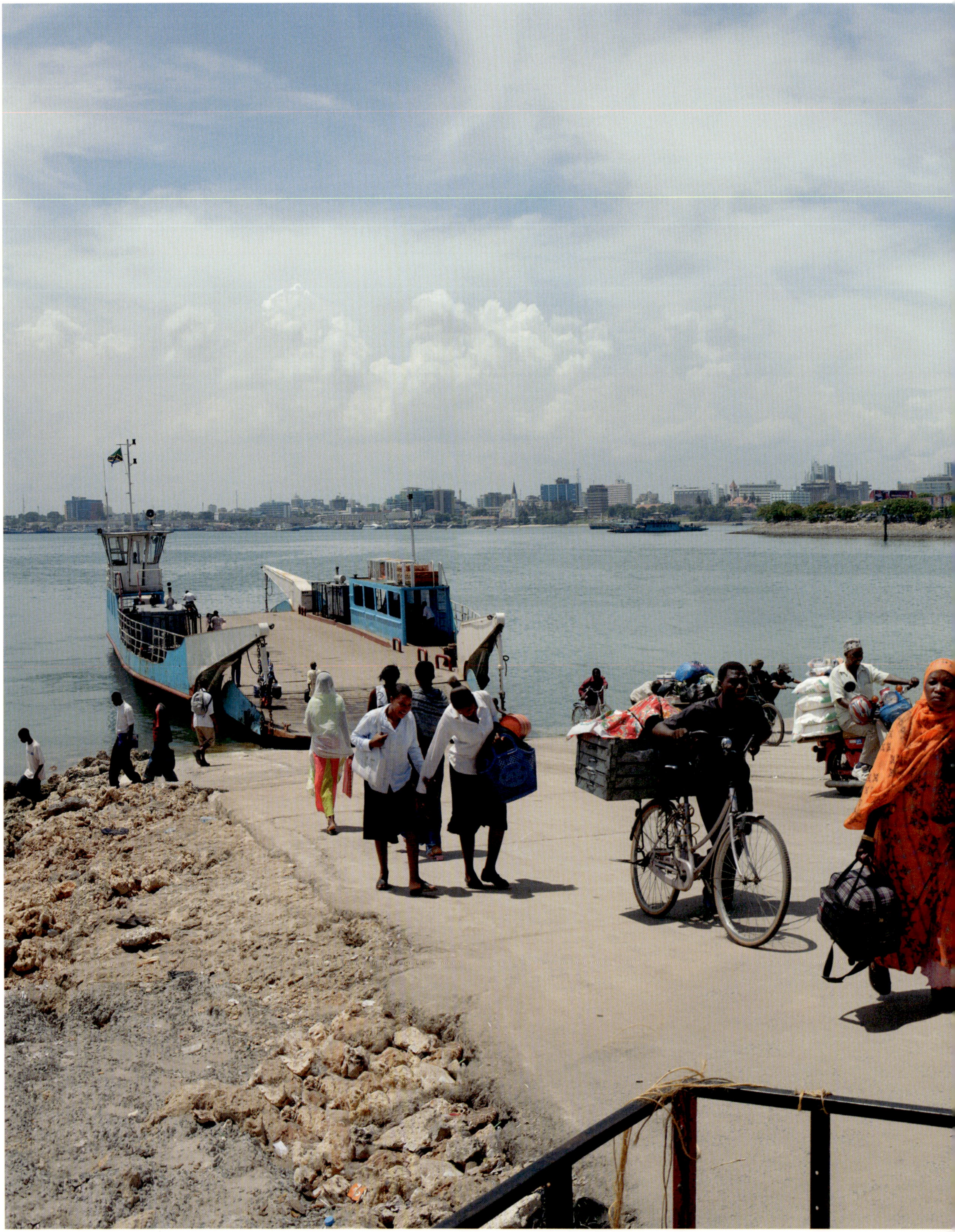

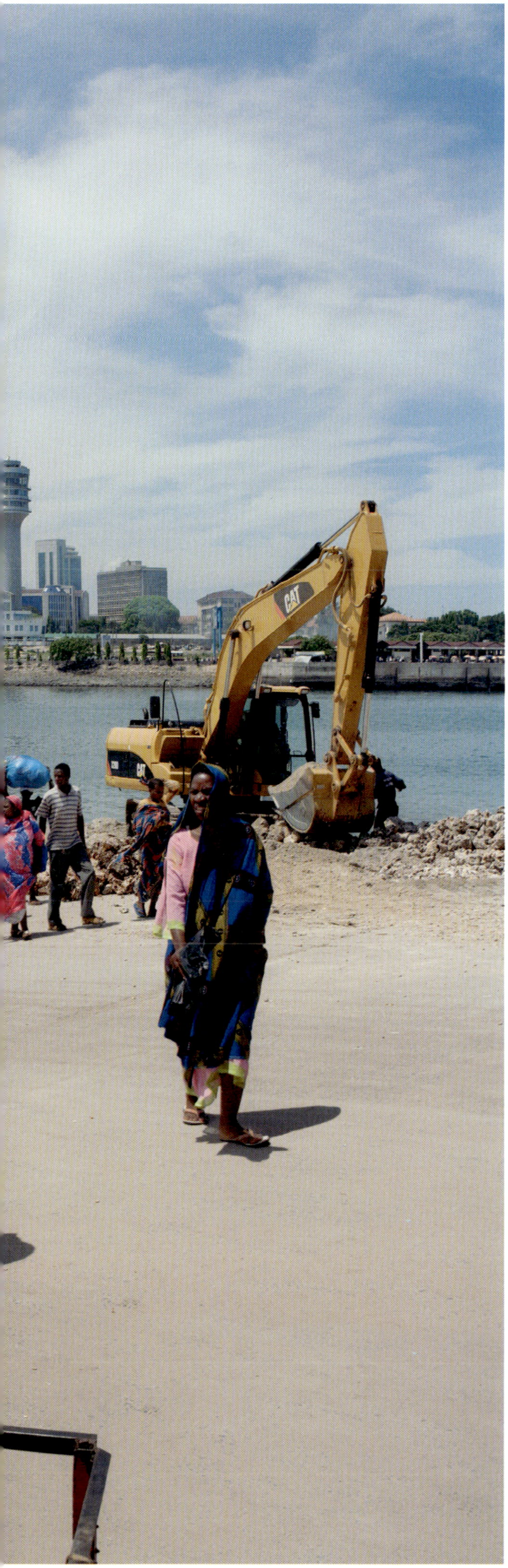

KIGAMBONI FERRY
DAR ES SALAAM

MOUNIRA MAIN STREET
CAIRO

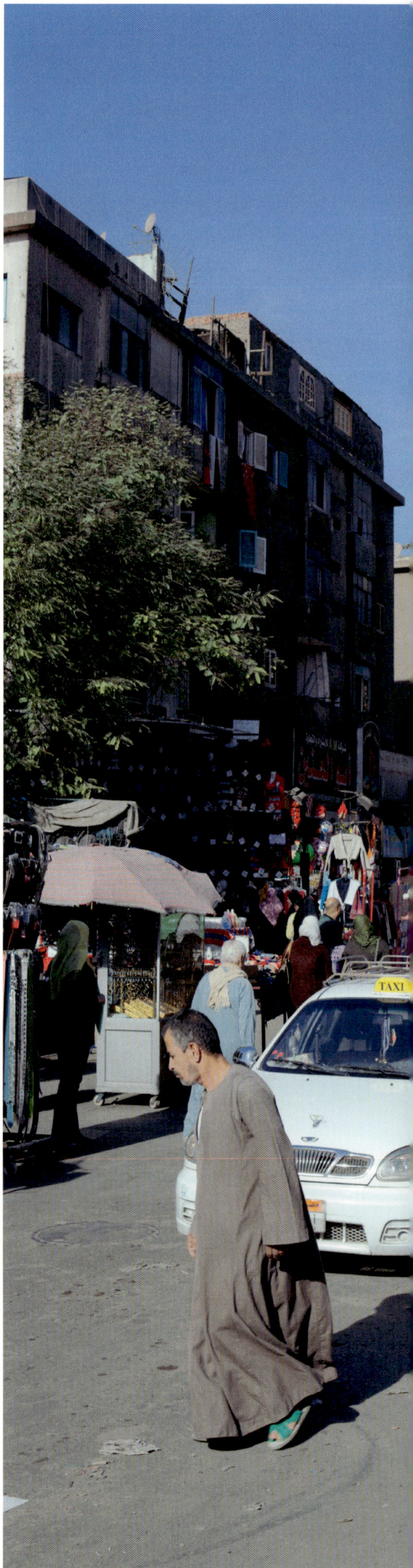

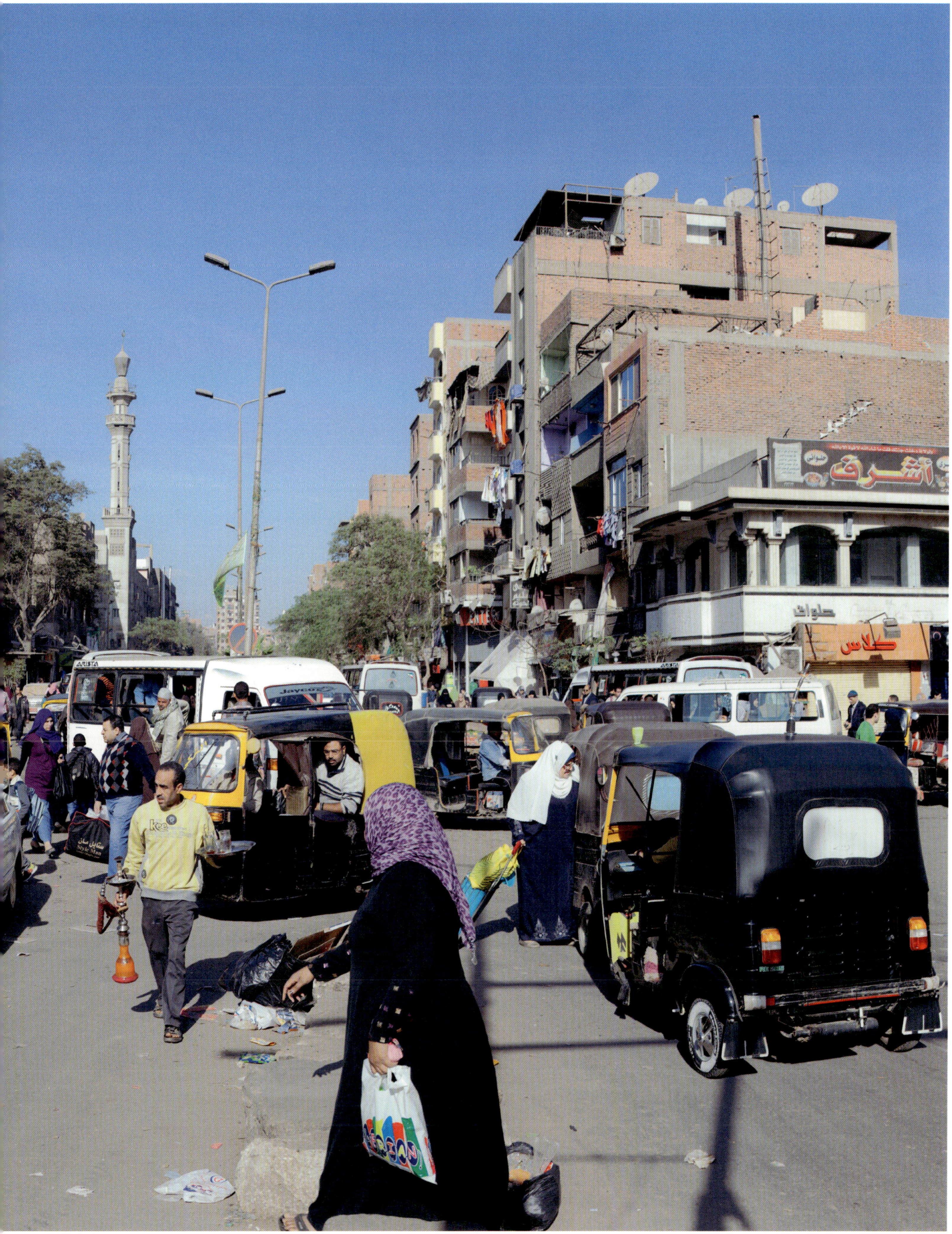
أشرف

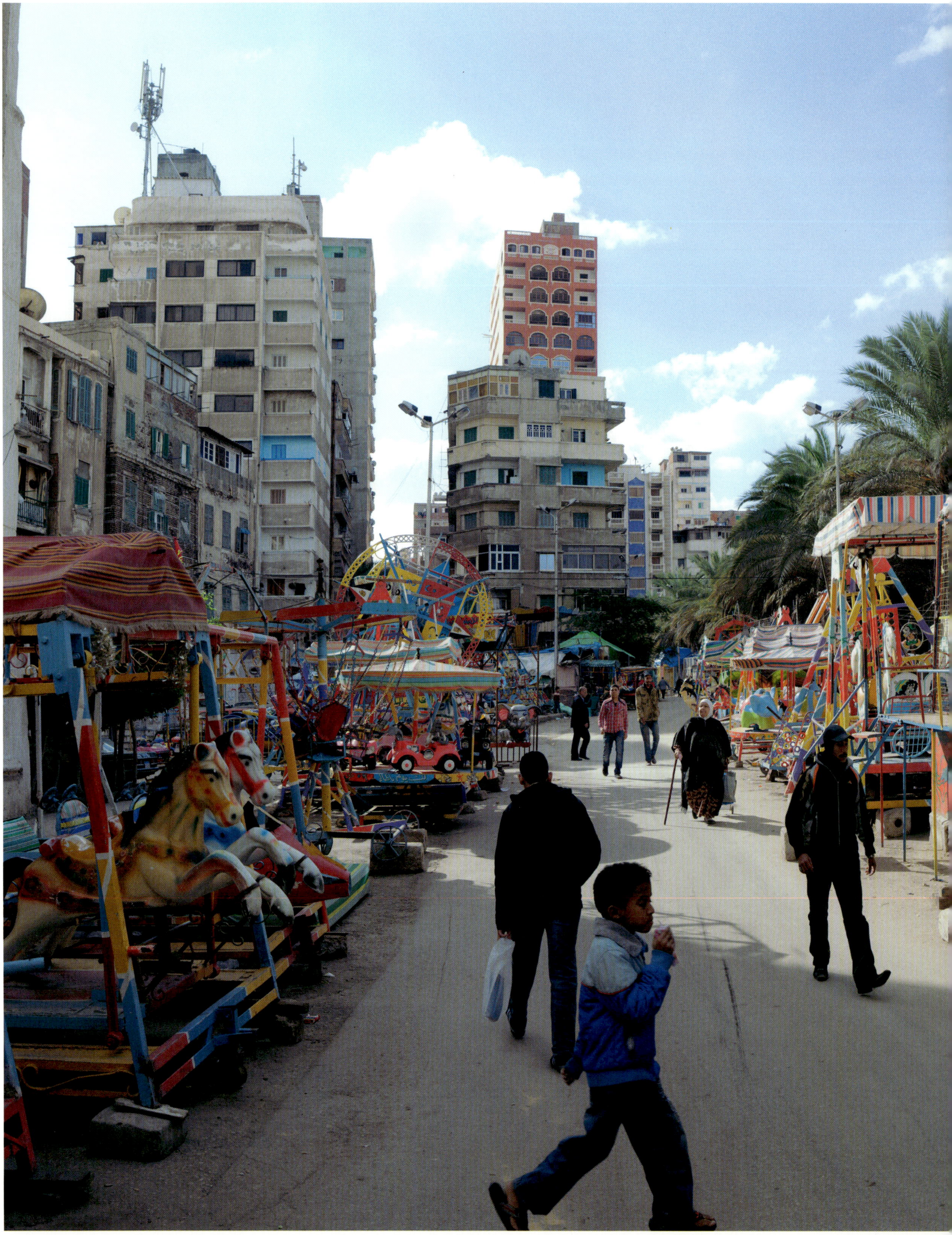

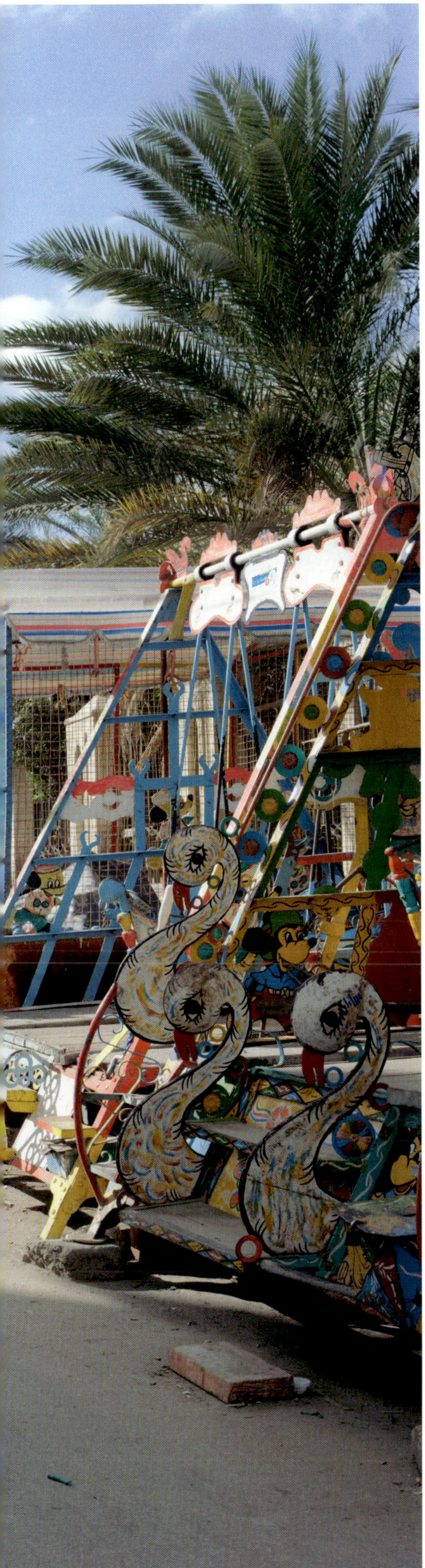

IBN ATTIA STREET
ALEXANDRIA

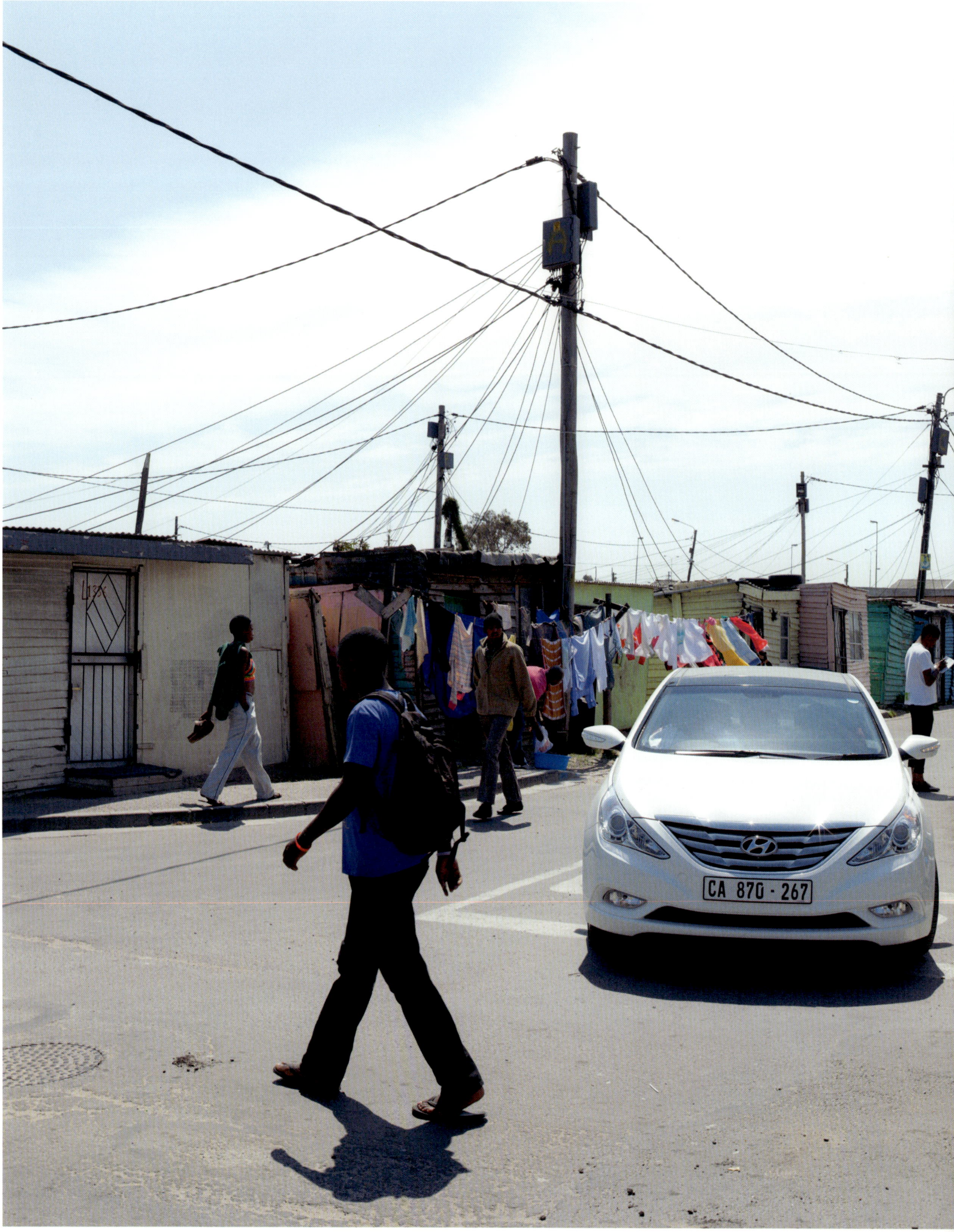
CA 870 - 267

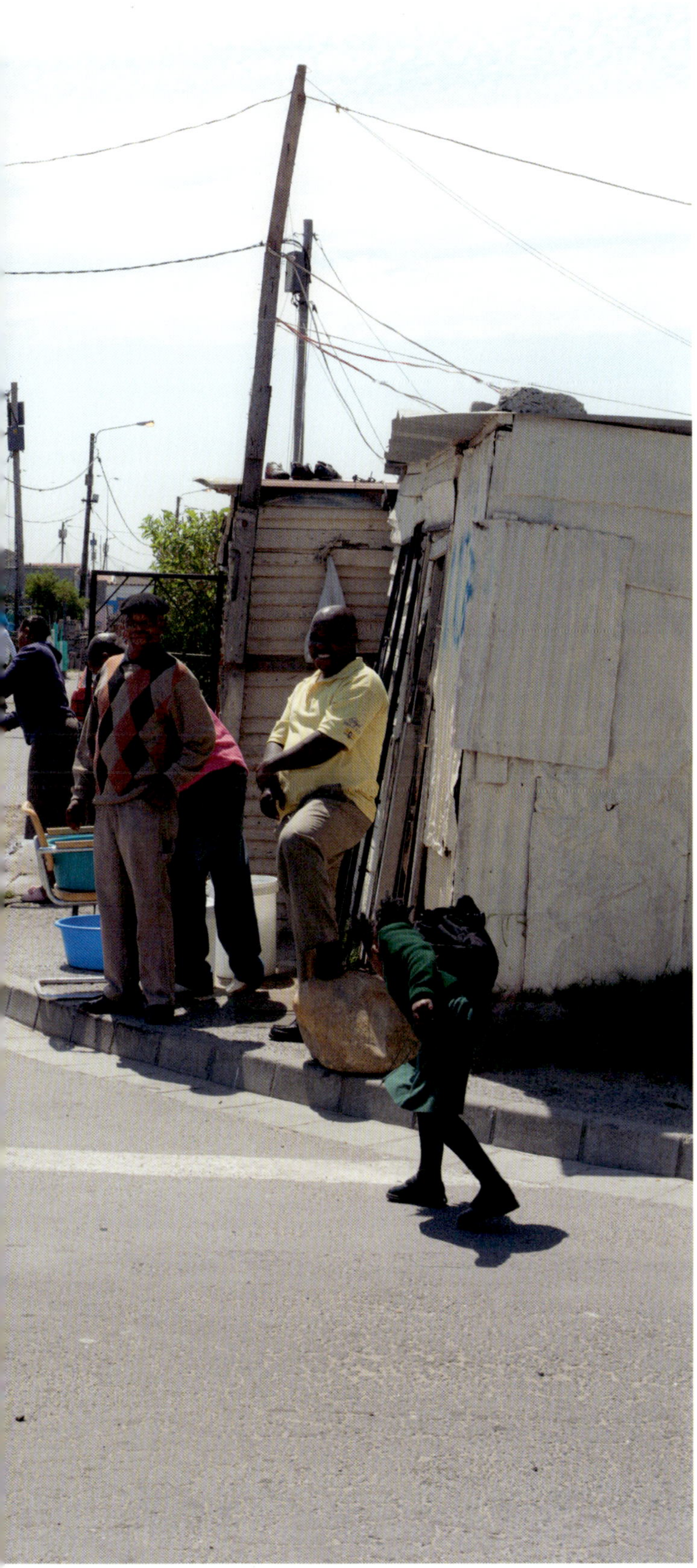

NJWELE STREET, K.T.C., NYANGA
CAPE TOWN

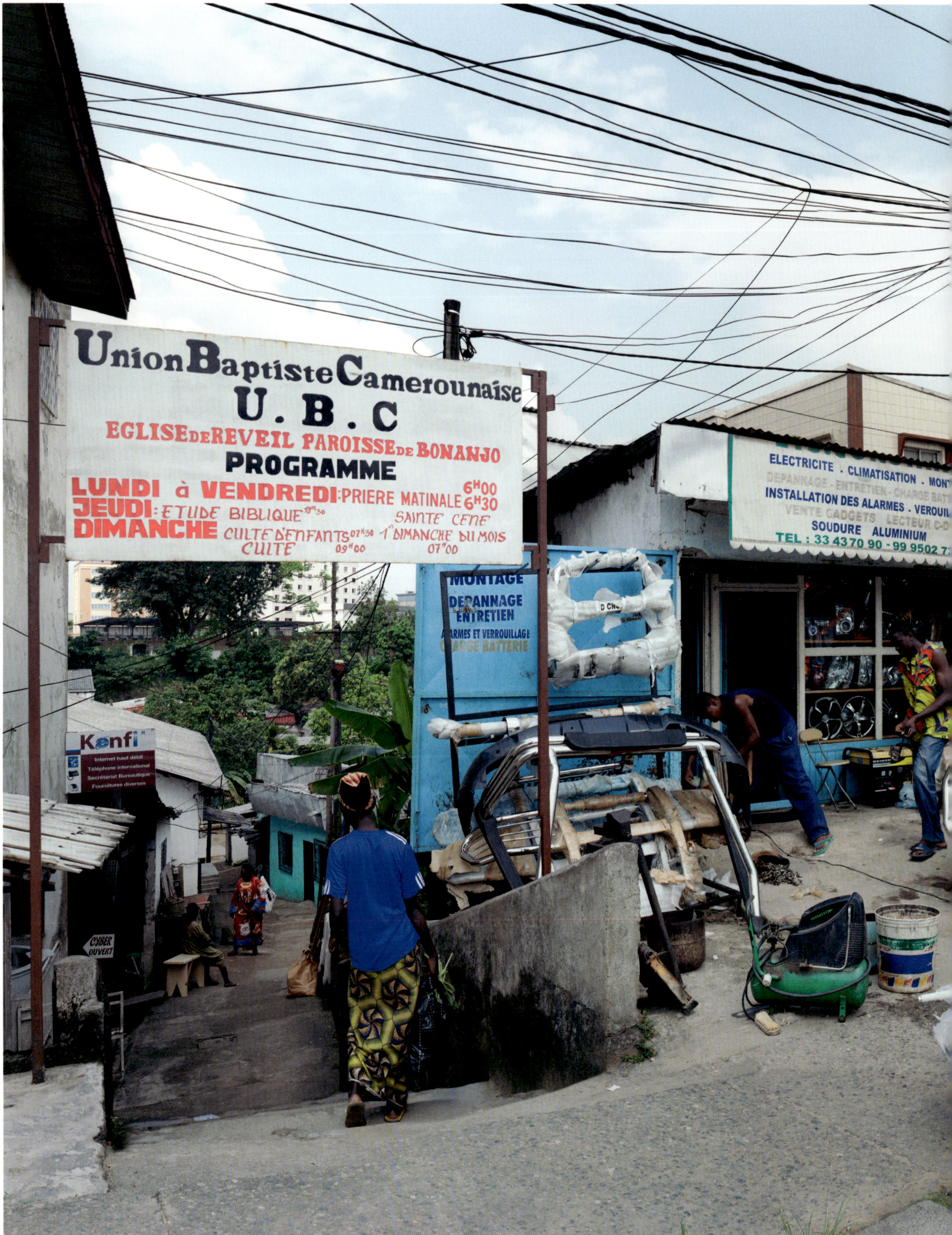
Union Baptiste Camerounaise
U. B. C
EGLISE DE REVEIL PAROISSE DE BONANJO
PROGRAMME
LUNDI à VENDREDI: PRIERE MATINALE 6H00 6H30
JEUDI: ETUDE BIBLIQUE
SAINTE CENE
DIMANCHE
CULTE D'ENFANTS
1er DIMANCHE DU MOIS
CULTE 09H00
07H00
ELECTRICITE . CLIMATISATION
INSTALLATION DES ALARMES
SOUDURE ALUMINIUM
MONTAGE
DEPANNAGE
ENTRETIEN
Kenfi
CYBER OUVERT

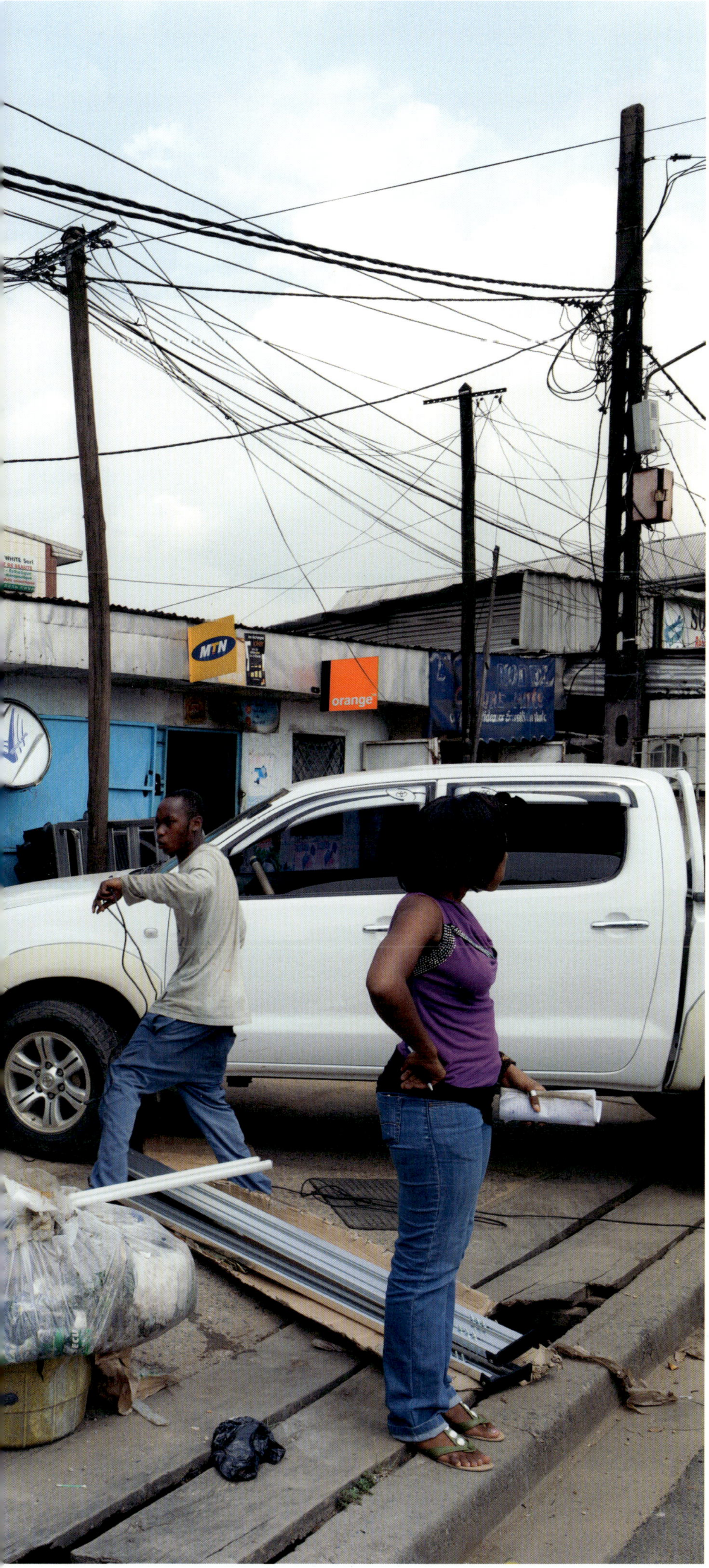

BOULEVARD DE LA RÉPUBLIQUE
DOUALA

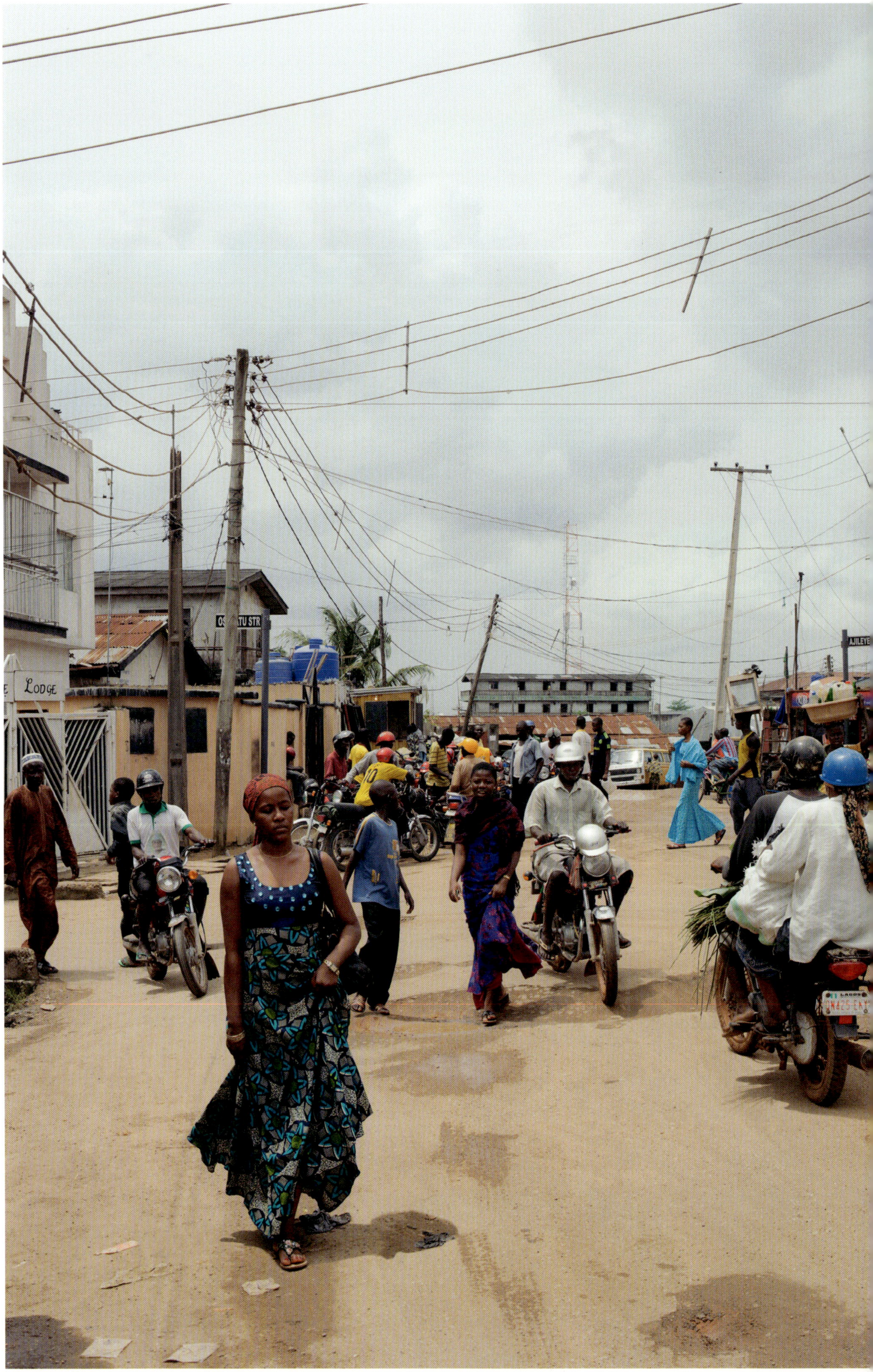
LODGE

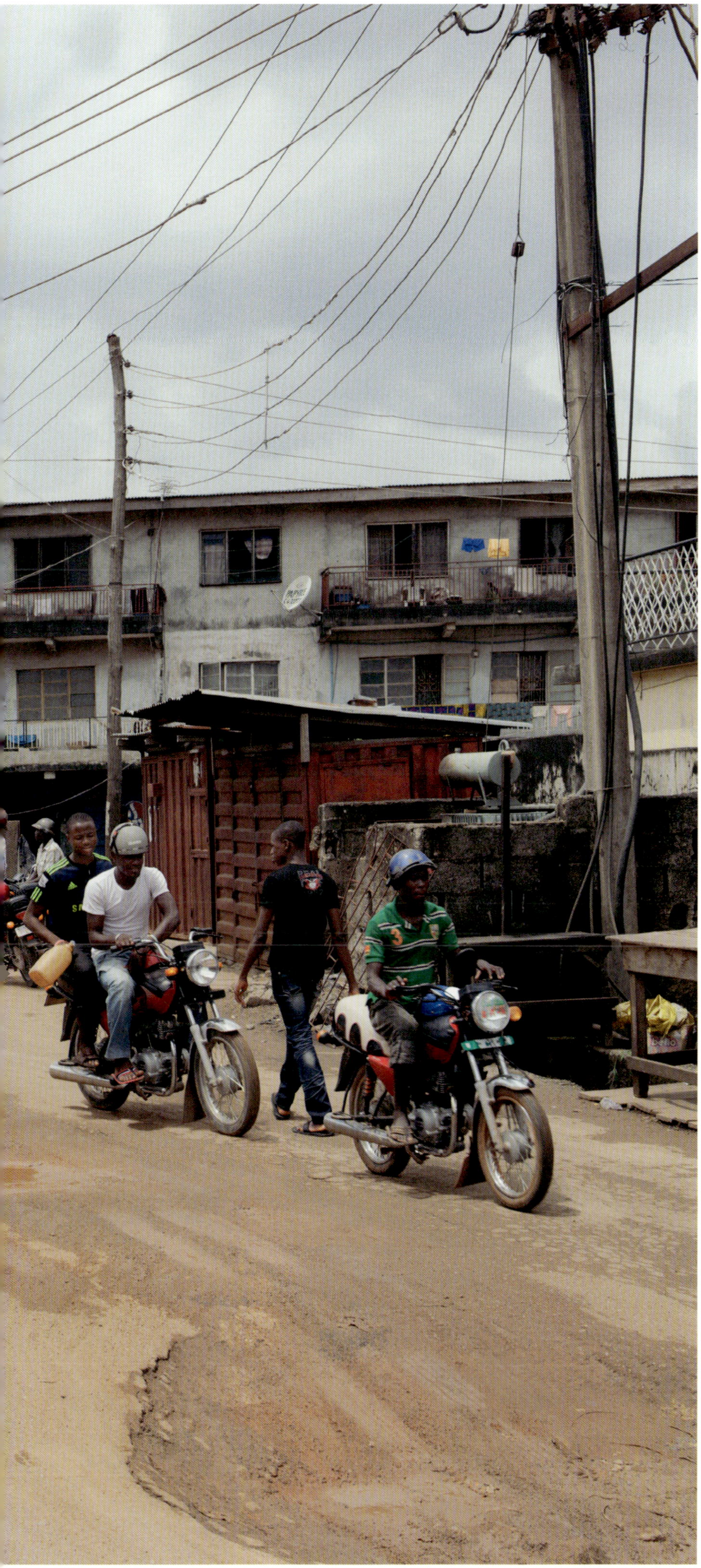

OREMEJI BY AJILEYE STREET, BRIGA
LAGOS

RUE DU ROI NJOYA, NEW BELL
DOUALA

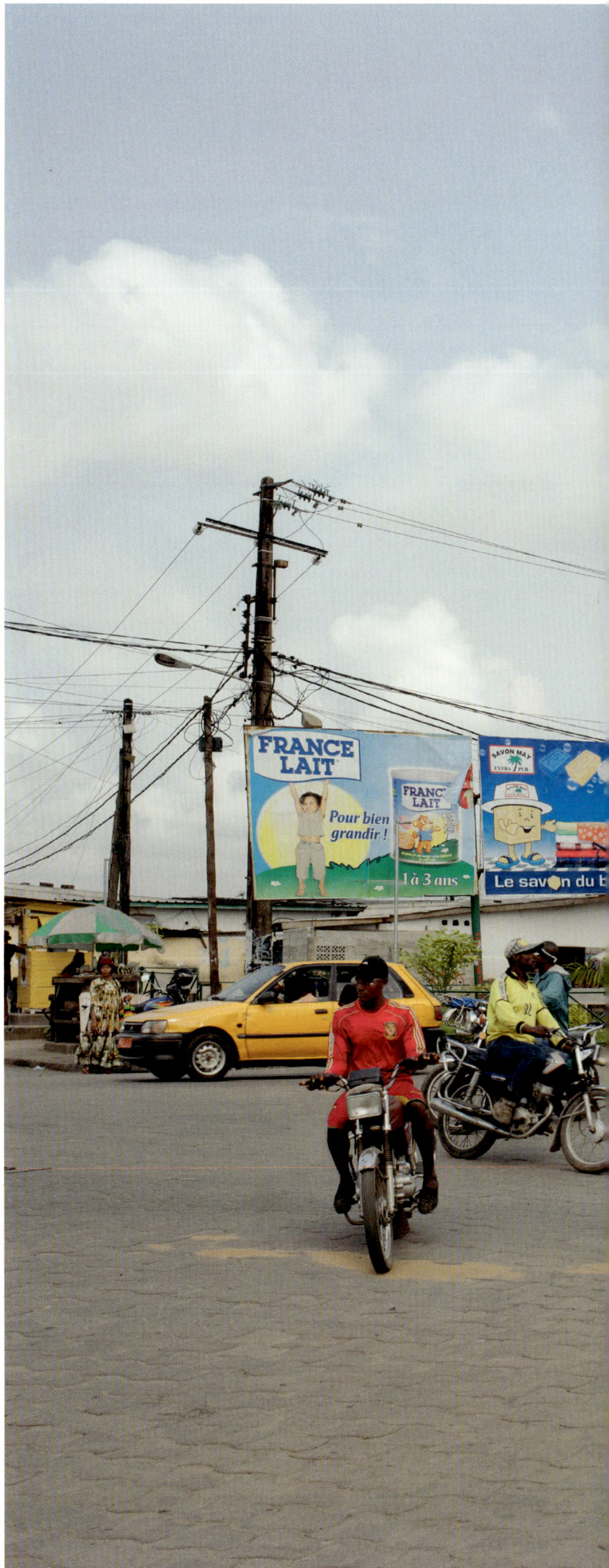

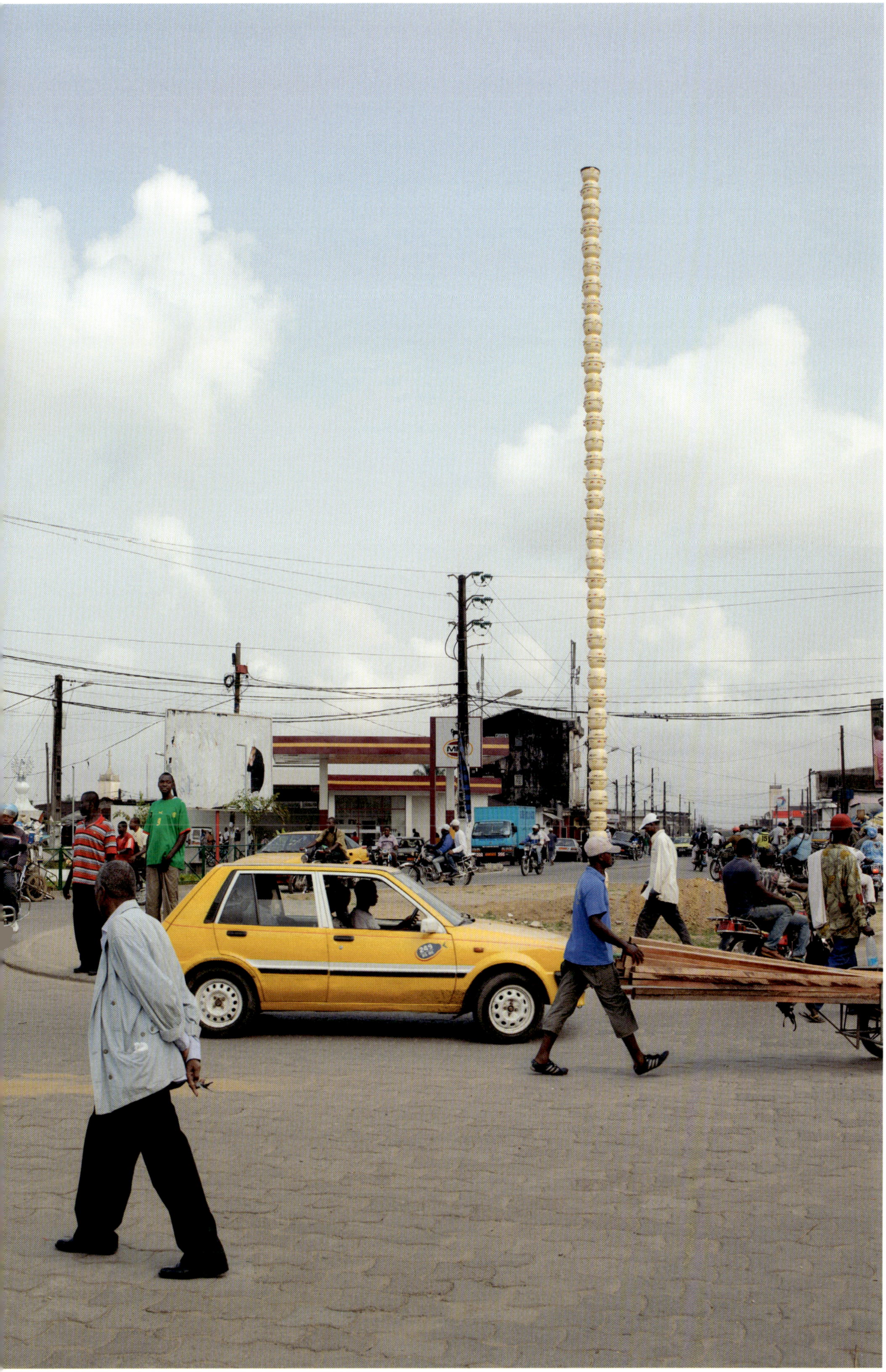

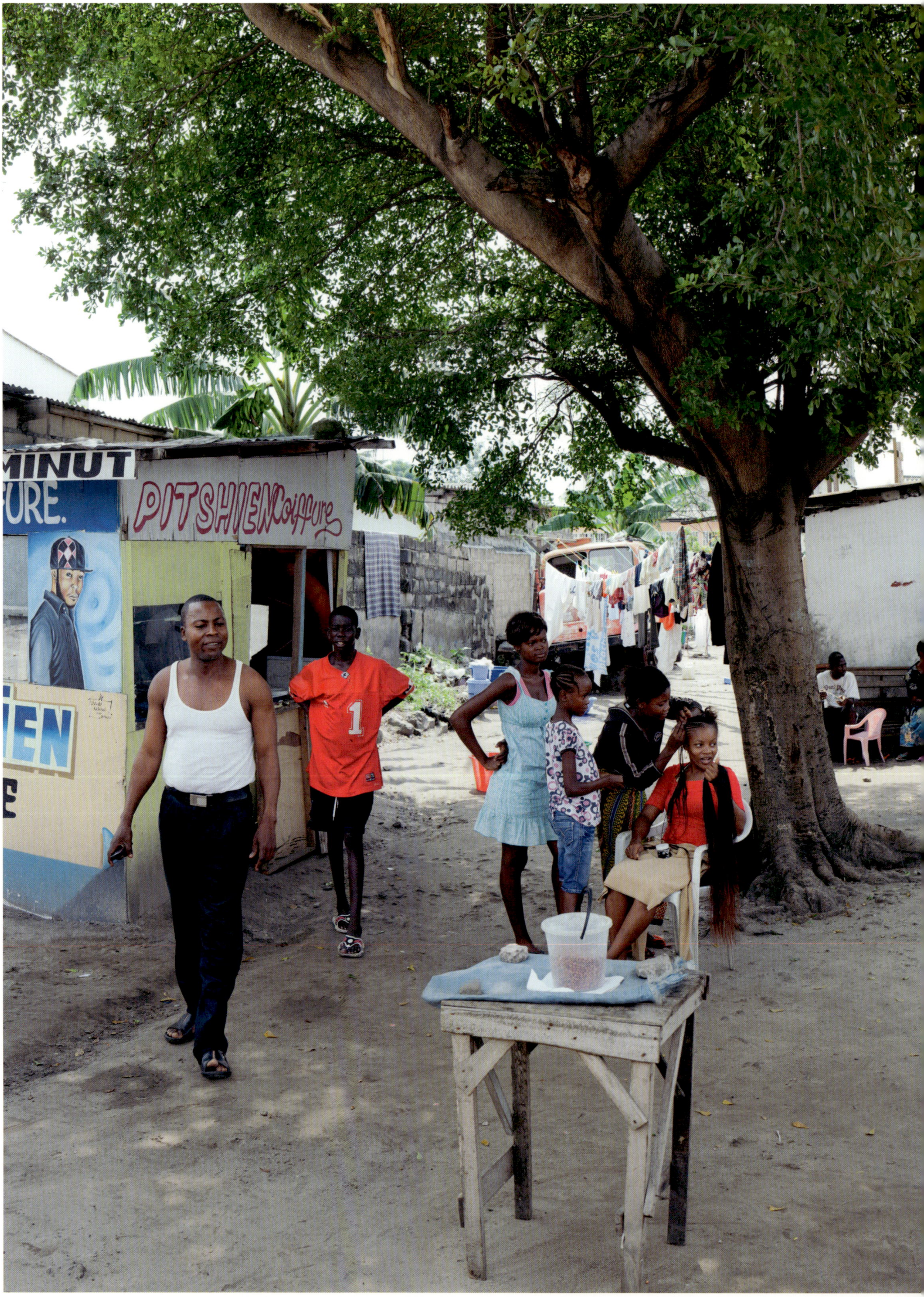
MINUT
URE.
PITSHIEN Coiffure
IEN

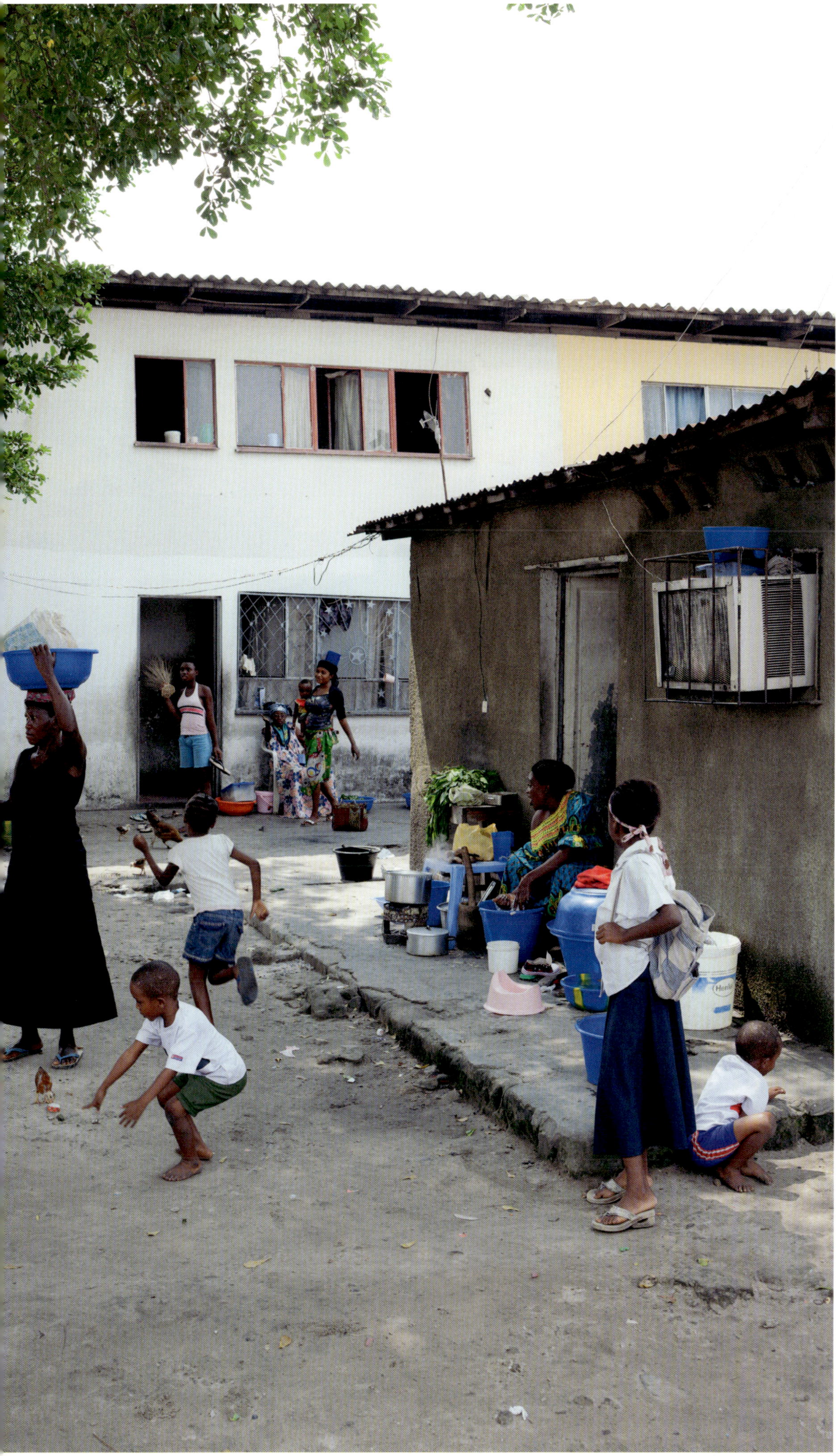

246 RUE MANZAWO, BANDAL
KINSHASA

Interview with Lard Buurman

CHRIS KEULEMANS

INVISIBLE CITIES

After traveling through China in 2004, where I concentrated on street life amidst the high-rises, I decided on Africa as my next subject. But there were no images of the African city in what I would call my collective memory. I had seen a documentary about Rem Koolhaas in Lagos, by the Dutch director Bregtje van der Haak, and Cape Town was a popular place to shoot commercials—that was about it. Westerners visiting Africa often pass through the cities on their way to nature and the countryside. During their stay in the city, they are advised not to wander into the street, which means that very few images are being brought back to the West. So I started reading about urban development in the global South and came upon Filip De Boeck's *Kinshasa: Tales of the Invisible City* in which he analyzes the urban infrastructure from the perspective of the human body. This brings about a different concept of urbanity. It might appear chaotic but it does work. At the grassroots level, urban space is being adapted to the local, daily, and often commonplace actions of its inhabitants. This triggered my imagination. My work is a portrait of everyday life at street level in the metropolis of today. Nowadays more than forty percent of the African population lives in one of the many urban agglomerations that have come into existence on the continent over the past decade. That means more than four hundred million people live in cities. Africa is urbanizing at a higher speed than the rest of the world. But how do these cities look? I read that half of Dar es Salaam consists of slums. To a lot of people a slum means open sewers, no paved roads, and people living in shacks. That seemed like hell on earth to me. And yet, there is life in these cities. What would happen if I were to continue my work there, taking my camera into the streets, both in the city centers and in the slums? Could I produce art that did not aim to only portray poverty and insist on solutions? Or would I have to admit that this was not my subject and return home with empty hands?

FIRST IMPRESSIONS

In November 2008, I traveled to Lusaka, the capital of Zambia, and Dar es Salaam in Tanzania, as the first stops on a trip along fourteen cities. And so my work on the African city began.

There are about 1.4 million people in Lusaka, but it looks like a provincial town. There are empty spaces and holes in the urban landscape. Traffic might get stuck here and there, but houses and buildings are mostly low-rise. I found out that the word "slum" can be misleading. There are slum areas, sprawl areas, and extension areas, which are informal settlements on the outskirts of the city. I was struck by how, within the same city zone, one place can feel more like a slum than the other. One of my guides, Mr. Mwanza, was in the process of figuring out how to make more of his plot of land by building more rooms so he could rent out more spaces—you could feel that kind of entrepreneurship all around you.

I had prepared myself well, taking medicine with me, but this was no jungle. There were shopping malls all around the city. It was the contrast between rich and poor that struck me the most. It is much more visible than in Europe, where cheap labor has disappeared from society, or rather it has been outsourced. The effects of globalization had confronted me in a way that it never had before.

SHED YOUR FEAR

In 2009, before departing on a trip that would take me through Johannesburg, Luanda, and Lagos, I was sure I was going to be robbed at least once. But it didn't happen—I have never been robbed in an African city.

On my first visit to Johannesburg my assistant Raymond, who has since become a good friend, told me where it would be safe for me to work—as long as I acted as a local and kept looking people in the eyes. The next morning I went out alone. I wanted to be able to move about freely.

But I spent the whole day afraid to take the camera out of my bag, and I ended up with just two pictures. One of them made it into my selection for this book (pp. 172–173). It shows a quite normal scene, not at all a dangerous place. Around 4 p.m., when the city center was emptying, I returned to the place where I was staying, physically and mentally exhausted. The next day, I went back. By taking these two pictures the day before I had shed my fear. Now, I was able to start working. My friend and colleague Hans Wilschut told me in advance: "Simply go out there and do your work, just don't be naïve."

For my first trips, I was worried that I hadn't planned everything meticulously, the way a professional photographer should. But it turns out that in Africa, one address and one contact person is enough. From there on, everything gets done easier than expected. I try to avoid hotels as they function as a border to the city. I prefer to stay at an art residency. I was able to get information as to where it is safe to walk, how to get a city map, and a cell phone card, and what the recognizable landmarks in the city are. This works better than arranging everything in advance. Often a local artist will be my assistant, or the residency helps me to find a fixer. My intuition guides me across the city map. I am often sent to the marketplace, but they are too obvious. I am more interested in the informal and unexpected places. I want to show the city in its full scope: the modernity and the business district, the fly-overs and the heavy traffic, but also the gray areas that merge into the sprawls. You can always recognize the sprawls on city maps because they are left blank, like a void. I am not aiming for the extreme: outright misery or exclusive wealth. I am just looking for life in the streets.

THE WHITE MAN'S GAZE

Many of the cities I visited were developed in the colonial era. The city center was meant for the whites, while their cheap labor forces lived in the periphery. The peripheries were not intended or designed for development. Lack of ownership is still an issue and makes it hard to create a future for your family. Why would you think about tomorrow, when there are no rules to create one. These places became (and still are) in-between places—not urban, and not rural. From the beginning, the economical infrastructure was set up to serve the motherland's interests. It was all about extracting resources: diamonds, gold, and slaves. All too often, this is still the case. Many African cities have to cope with the problem that there is no internal market for its natural resources; those are turned into valuable goods elsewhere. Thanks to Congolese coltan we can produce our smartphones, but back in Kinshasa there is no food on the table.

Gradually, I became more conscious of what it means to be a white man with a camera in Africa. Walking around in these cities, I try to aim my camera beyond the cliché, beyond the sense of pity, and the conscious or unconscious guilt. I am not trying to prove to viewers that Africans live in real cities too. My photography doesn't claim to make bold statements. In Johannesburg I did a shoot in a park with a lot of Zimbabwean refugees (pp. 102–103). When you don't know the situation you can interpret it as people enjoying leisure time in the sun. And partly, this might be true as well. My aim was not to shoot the most harrowing images. I was there to show something else, the daily life that often remains ignored. That people live on the margins doesn't necessarily mean that their life is pure hell. It may very well be the normal situation there and there is not much I can do to solve that, but I can learn a lot from it.

It's important for us in the West to keep looking very carefully. I believe it's good for a society to have both perspectives, from the inside and from the outside. These times of globalization turn us into people that are afraid or overcome by fear, at the very moment that we are enjoying the benefits. That same fear causes us to make an attempt at keeping African people out of Europe, out of our aging continent, or at best turning them into second-rate citizens. All the while we need the younger African populations to support the European economy. Taking into consideration globalization and the resulting diaspora, I believe European cities will become more and more similar to African cities in the future.

POSSIBILITIES

The urbanist and sociologist AbdouMaliq Simone looks at the African city as an infrastructure that doesn't resemble what we know from Western cities but functions nonetheless at a grassroots level. His theories have inspired me and have provided me with a sense and motivation to this work. There is value in looking at everyday life that takes place at street level. We should not neglect the space where we, as a society, meet each other. If we would all live a gated life, we would all think gated. I am not attracted to gated neighborhoods because that is not where you will lose your fear. When you consider everything outside your cocoon to be unsafe, be it your house or your car, the city will never belong to you. When we come to Africa it is vital to live and look outside of those cocoons, because there is a great freedom in being able to walk unprotected in public space.

I learned to see normality in what at first seemed strange to me. Take Luanda or Lagos. The informal economy is a given. Goods are literally being carried to you from all sides. When you're in a taxi, stuck in a traffic jam, someone will approach you with, say, a propeller fan. It might not seem practical at the moment, but you get used to it. You don't need to head to a supermarket, but you can start choosing from what is spontaneously offered to you. And the guy with the propeller fan is just doing his job. A traffic jam can be irritating, but it can also turn into an advantage.

In Douala, Cameroon, I took a picture of a four-wheel drive parked on a sidewalk (pp. 140–141). The street had been turned into a workspace and the sidewalk into a shop, but a shop that you can walk right through. Public and private space overlap.

The same goes for the pedestrian bridge on the cover of this book. The chair this man is carrying could belong to him, but he might as well be a courier. Those ladies carrying baskets might be heading out to sell food. There is a great wealth of possibilities going on in these streets. And the girl having her hair done (pp. 146–147): Is the other girl her friend, or her hairdresser? Is this a mother cooking for her children or is she running a street restaurant?

The lack of infrastructure looks chaotic at first, but the apparent chaos makes room for the unexpected. In our part of the world, we tend to control space and play it safe more and more, which means that we have more to lose and that many things are no longer possible. I do not want to romanticize the possibilities inside this chaos. I just want a wider perspective by forcing myself and the viewer to look beyond our first impressions of a place, and to make contact with people—one that is not based on the intention to help. Experience shows that much of what we have tried to do in these regions has ignored their true needs. From a Western urban perspective, how do we know what's good for Lagos—a city of thirteen million people, incomparable to a European capital.

The public space in African cities is characterized by the absence of infrastructure, but also by liveliness, possibilities, and space. The shifting borders between public and private, formal and informal, are striking. There is so much to investigate because there is so much life being lived in the streets. All the physicality going on in a public space is far more diverse than I have ever seen in other cities. So in practice, I have become even more of a documentary photographer—in an informal way. In my African photographs the choices in composing the scene afterwards became even more subtle when compared to my series in Europe and China. This is also due to a development in my work, I am more daring in finding the ordinary—to show the way things normally are. Walking around with a camera, I am fairly visible. People approach me, respond to me, they are curious about my presence and why I'm taking pictures. That doesn't always help me do my work. If you look at this picture taken in a dusty Lusaka extension (pp. 194–195), there were kids laughing and dancing right in front of me almost the entire time. But I want to create an image that it is not about what I cause as a photographer, but about what is happening when I'm not there. So I adjusted the scene afterwards and created a much quieter image of men and boys walking away from me. A photograph that was actually impossible for me to shoot. And yet, it shows reality all the same.

THE PEOPLE

Sometimes, people object to me taking photographs. I've learned that people may object for many reasons. Some have religious objections and sometimes it's about money. In Cairo there had been warnings on television that Western photographers are often spies. In Kinshasa, everyone kept asking me for my permits, not just the police. They live in what I would call a post-corrupt society. The corruption and bureaucracy has pervaded everything to such a degree that it starts creating rules of its own. People would gather around me in large groups and were verbally but not physically aggressive. They would start discussing my situation: did he receive his permit from the municipality and not from the state government, or the other way around—endless discussions that somehow opened up a space for negotiation.

I was "arrested" by the student security at Kinshasa's U.P.N. Université—my assistant and I hadn't realized that we had entered private property. After negotiating the fine, I was surprised by the head of the security group when he said: "So, where do you wanna make pictures now?" He escorted us around the university for another two hours. I respect the fact that people don't want me to do my work around them. Sometimes I can explain what my aim is through my local assistants. I always explain my interest in daily life.

In any case, my work is about the individual in the masses, about the human body in the crowd. Not about the mass itself. Football stadiums and marketplaces turn all the individual people into one huge body, a shapeless crowd. I am trying to present the relationship between public space and the shifting collection of individuals. It is my hope that the work entices you to wander through the image with all the different people, their posture and their position in space. These people are all worth watching—a variety of people, all deserving a closer look. Surrounded by the multitude, consciously or not, they stand out. This is the contradiction of life in the city: having outgrown the village, you have to become more individual—alert and outspoken—but on the other hand, in the city you are just one among many, less visible than you used to be in the countryside. In the urban public space, there is always a paradox at work: everyone passing you by is part of this anonymous mass, but the moment they walk by, it's just me and this person, one on one. In that moment, we have something in common.

There is of course a great variety among the African cities I visited. An Islamic city is different from a Christian one. A city oriented towards the West is different from one that turns the other way. They have their own urban styles. I have seen slum dwellers go to work in immaculate white shirts, or young guys in spotless hip-hop outfits. This sense of dignity is on display in the most unexpected places. I had wondered why a person in a slum would buy an expensive car or a satellite dish. Why not move to a better neighborhood? But in the African urban context you might lose the connection to the network you belong to. That's why many people don't move from the slums, even if they have the chance, because this invisible network is what carries them.

CREDIBILITY

I often construct my photographs, selecting from many moments shot on one location, in order to create a sense of scene. But I don't like to exaggerate. I am not out to prove that a city is always crowded. By not zooming in on a specific person, I am playing with movement in space. I hover on the edge between a staged situation and reality. If I were to only give people instructions or direct them the way I do in my commercial work, my photographs would lose their credibility. And to me, credibility is the essence of documentary photography. And yes, playing with the composition, crafting the image, making choices about who to show and who not to show, all of that can very well be compatible with credibility. Even though we live in a time when we all know that a photograph is not reality itself, people still tend to believe more strongly in images than in written text. But every photograph has been manipulated! The frozen moment itself is a construct of photography.

My method of working might offer a more realistic view of the African city, because of my intention to do so. I believe my work presents a quite nuanced image of the urban space. Although not pretending to show truth, I might actually be approaching it very closely.

JUNCTION #3

CITY OF FLUX

"A real city comes alive when people flood its veins and arteries like so many ants in a complex nest. A real city is a musical instrument. It is meant to be interactive. To have people sing through its strings and hollowed bones.' *

* Chris Abani, "This City," pp. 73–83

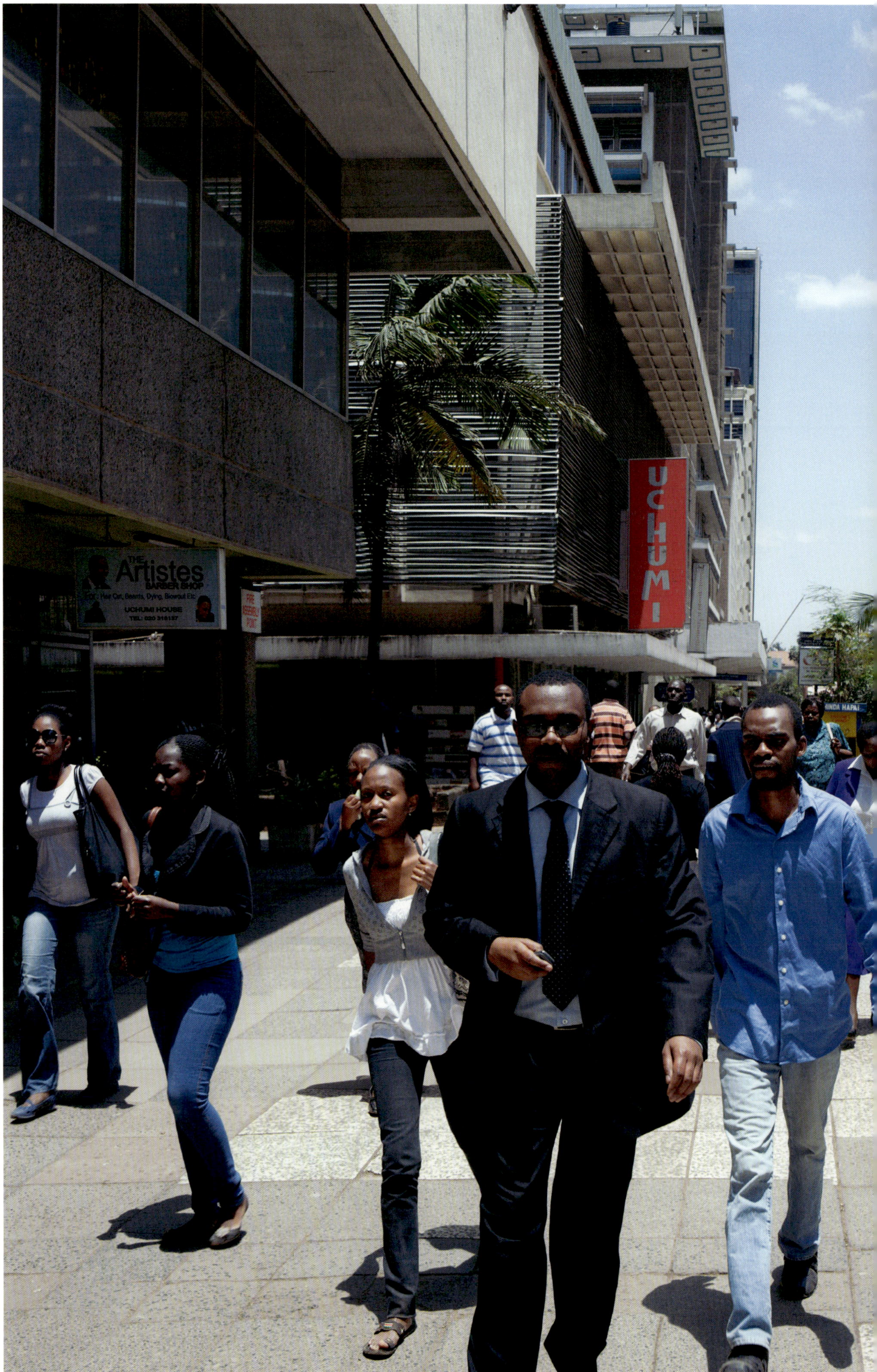

AGHA KAHN WALK
NAIROBI

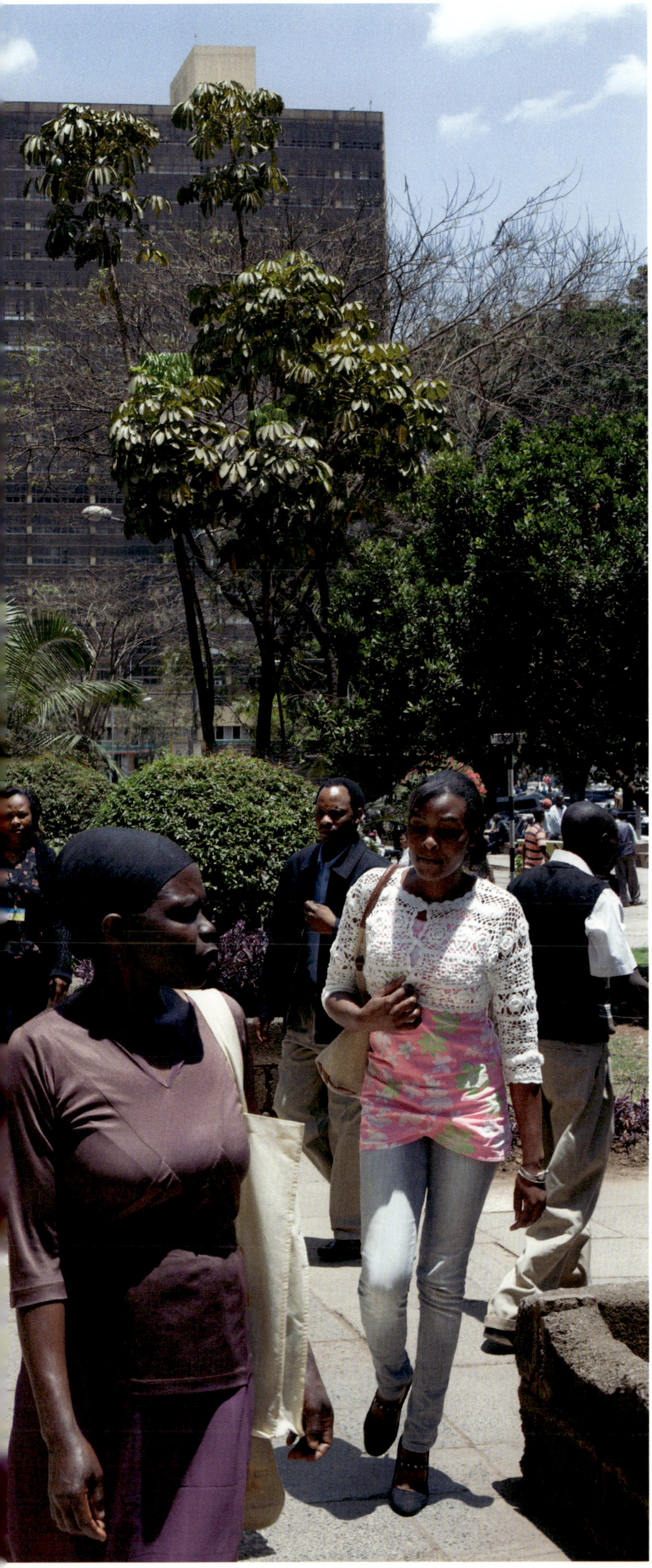

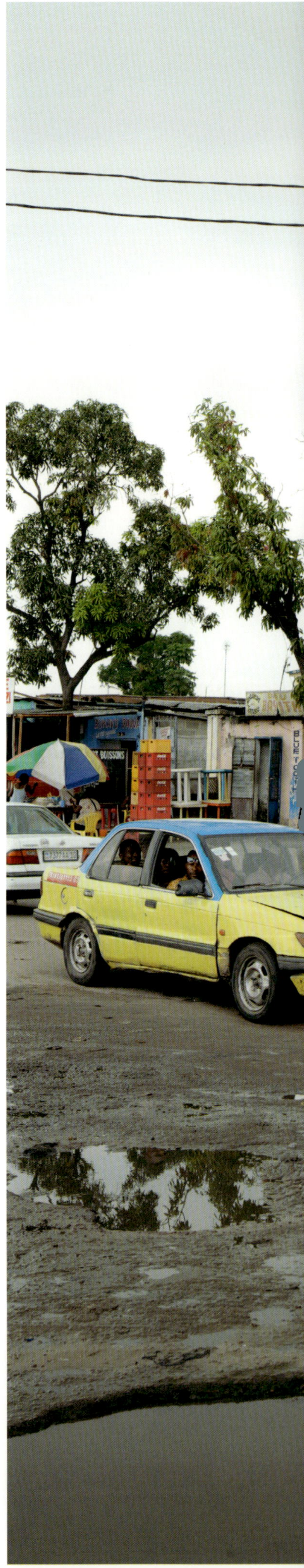

AVENUE SAIDI
KINSHASA

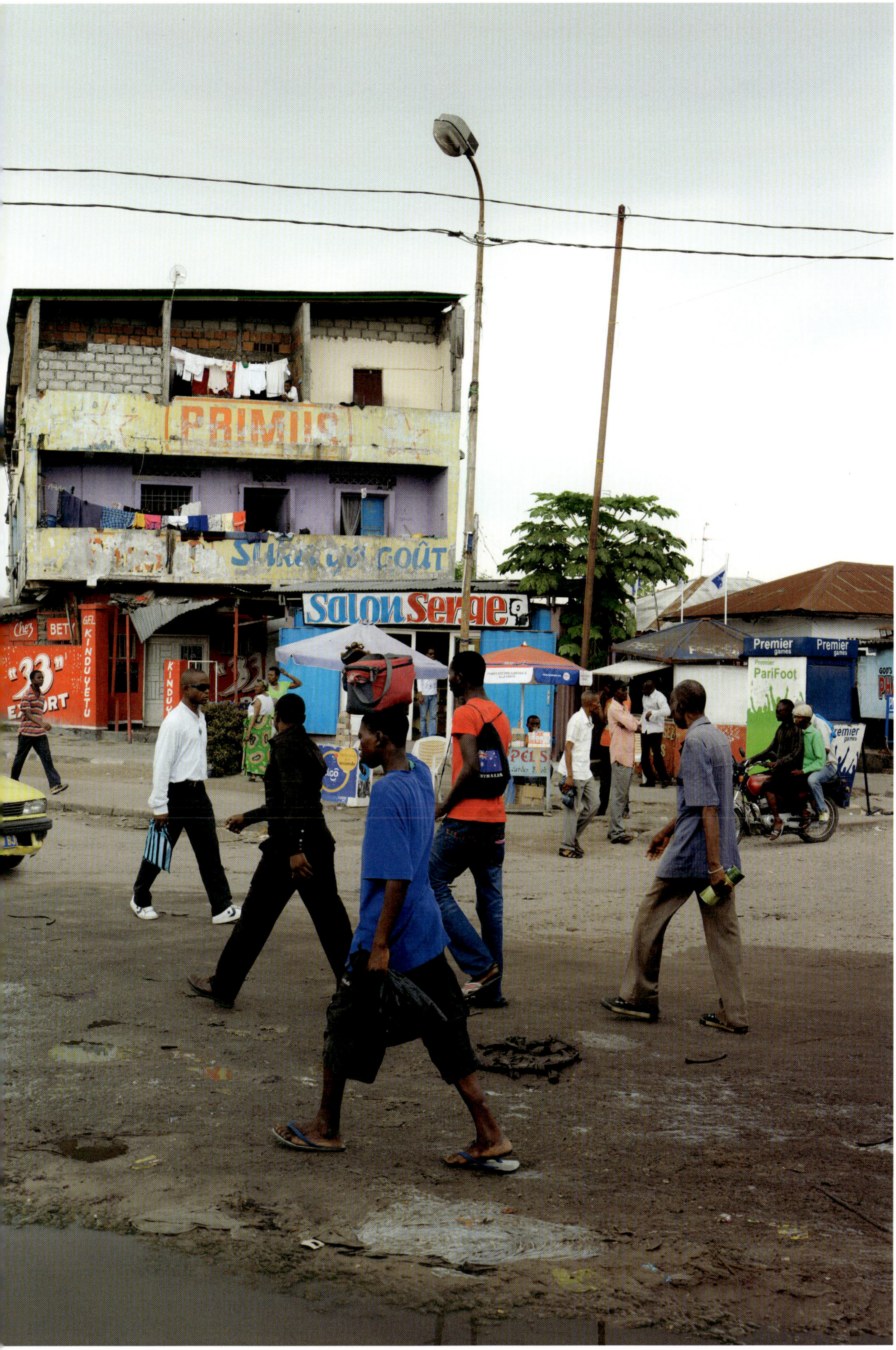
PRIMUS
Salon Serge
Premier
Premier
ParIFoot
KINDUVUETU

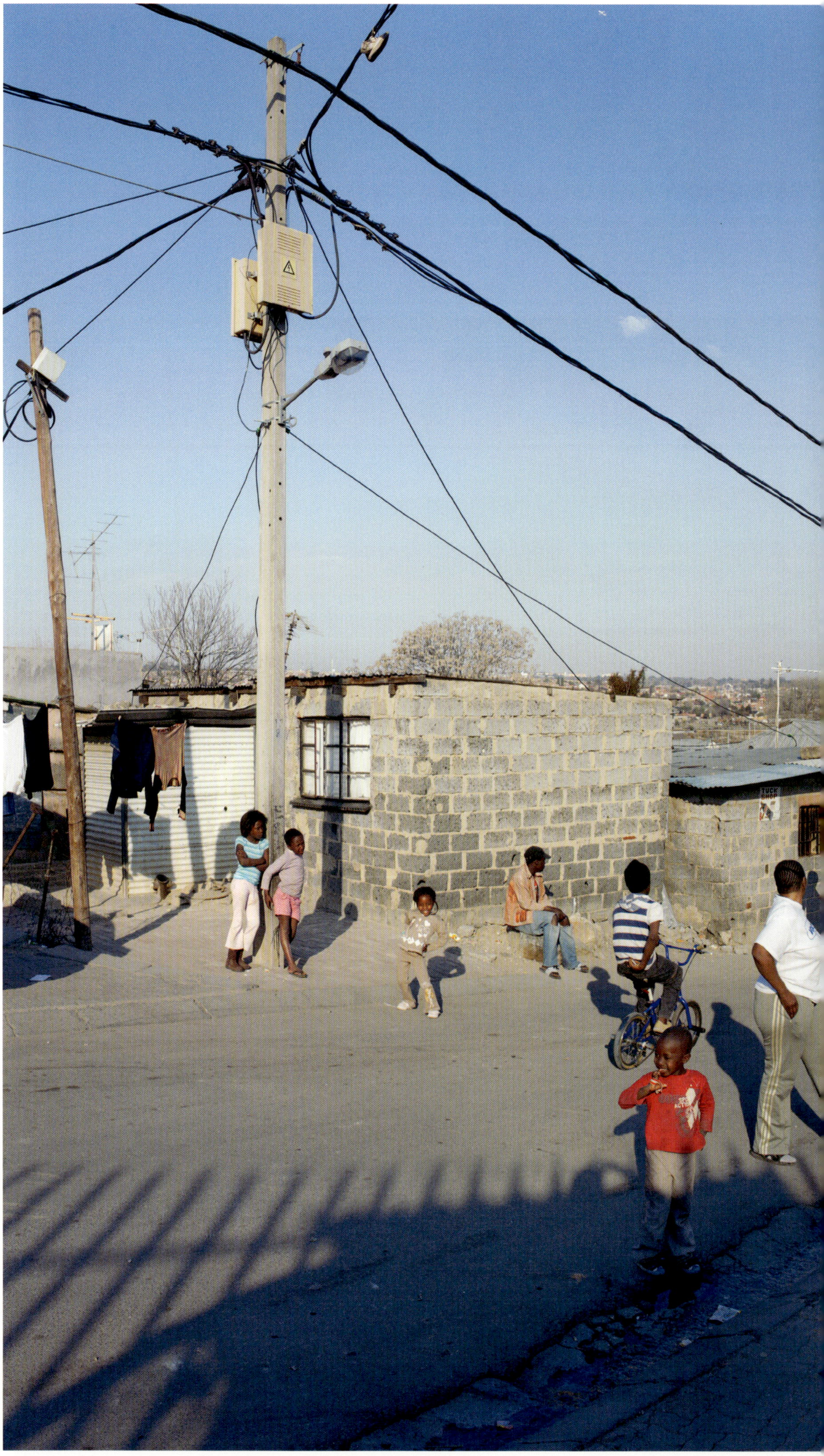

17TH AVENUE, ALEXANDRA
JOHANNESBURG

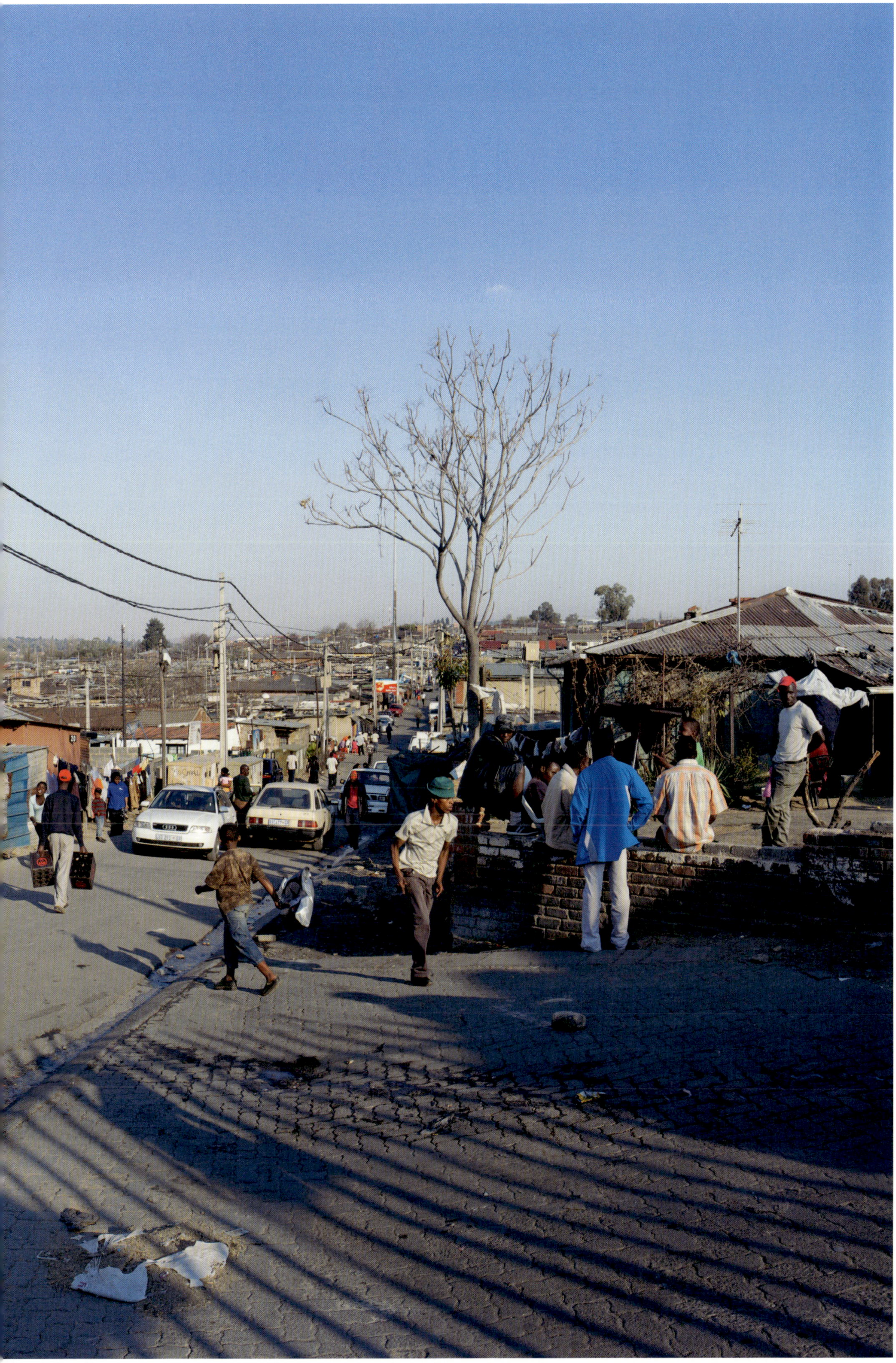

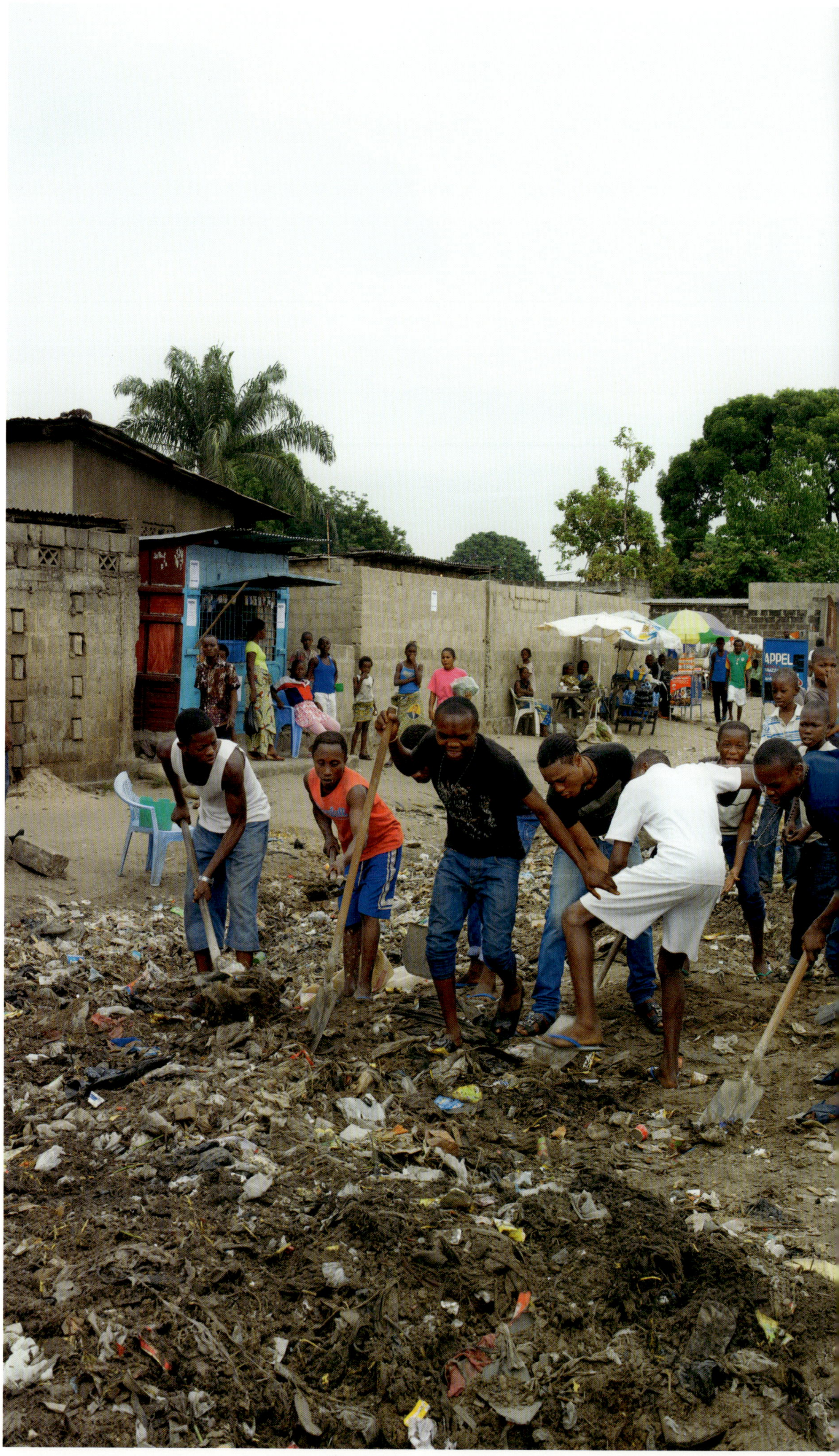

AVENUE ZONGO, BUMBU
KINSHASA

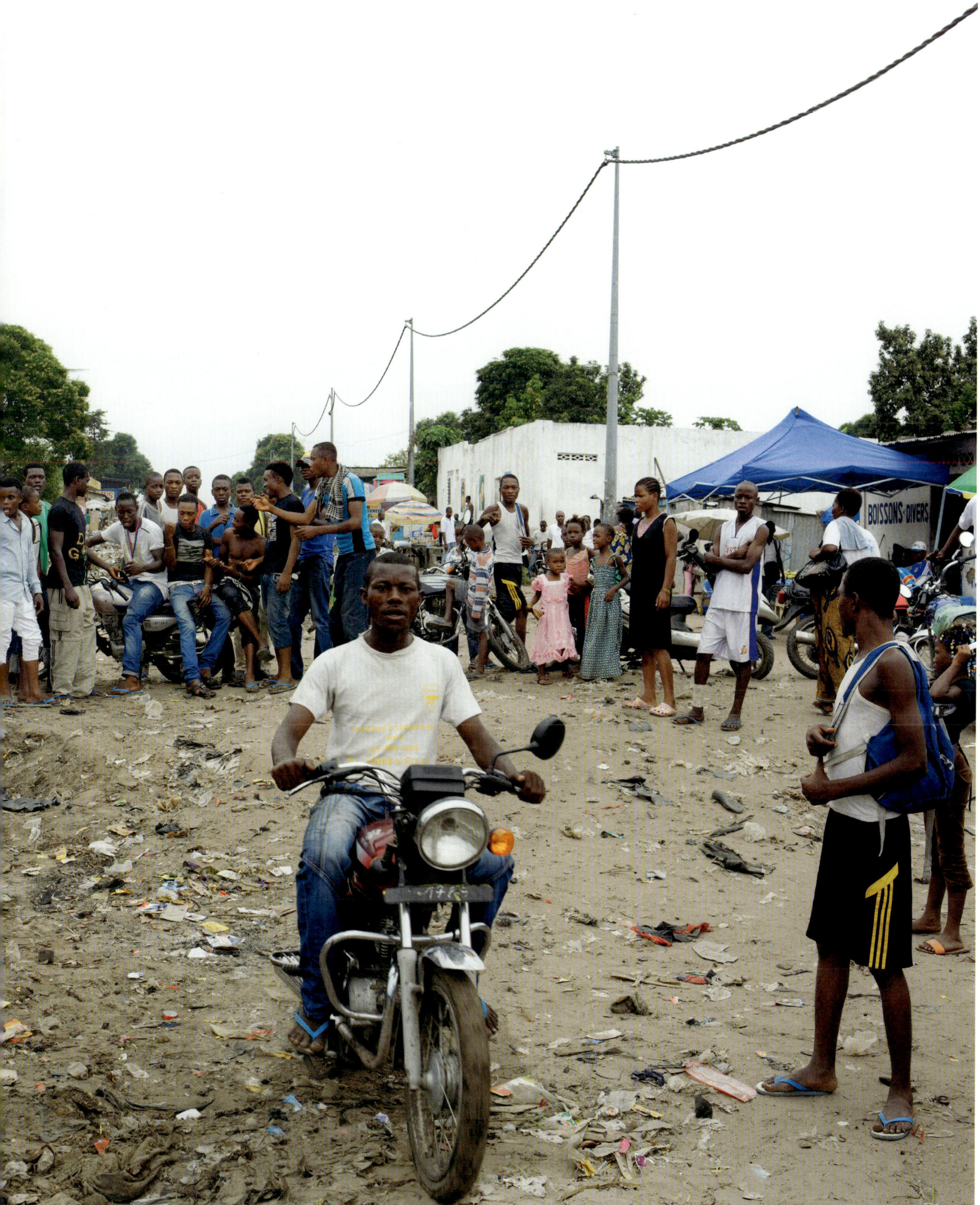
BOISSONS-DIVERS

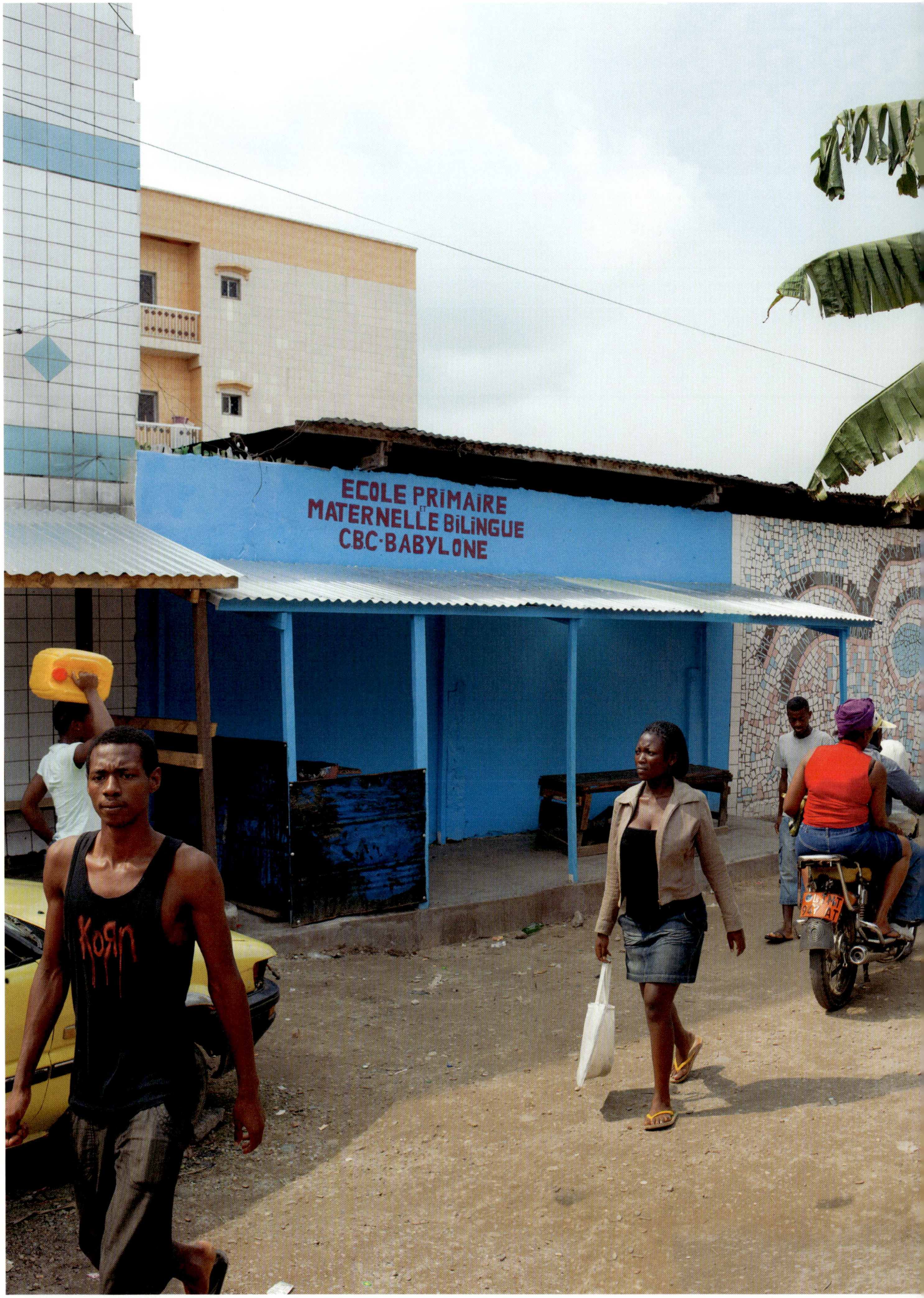
ECOLE PRIMAIRE
ET
MATERNELLE BILINGUE
CBC-BABYLONE

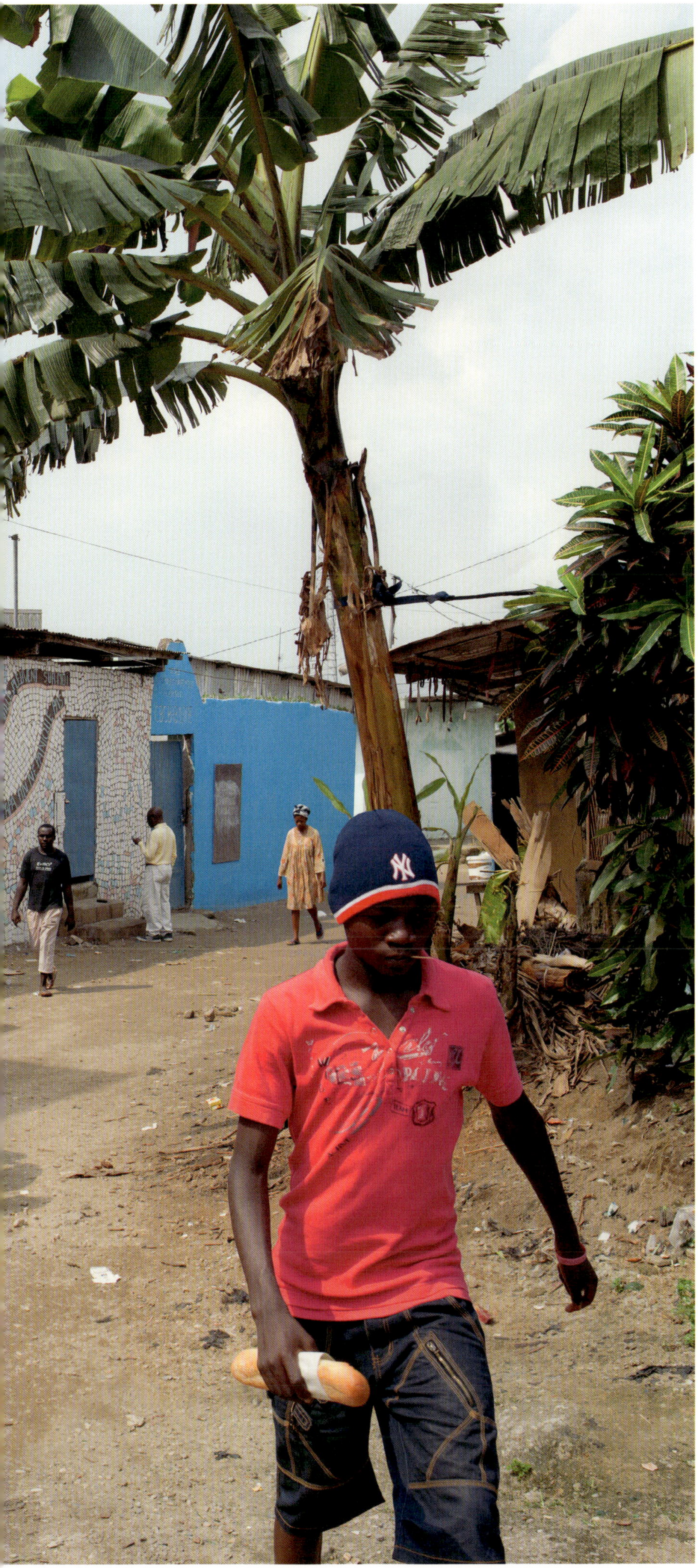

C.B.C. BABYLONE, NEW BELL
DOUALA

KIVUKONI FRONT
DAR ES SALAAM

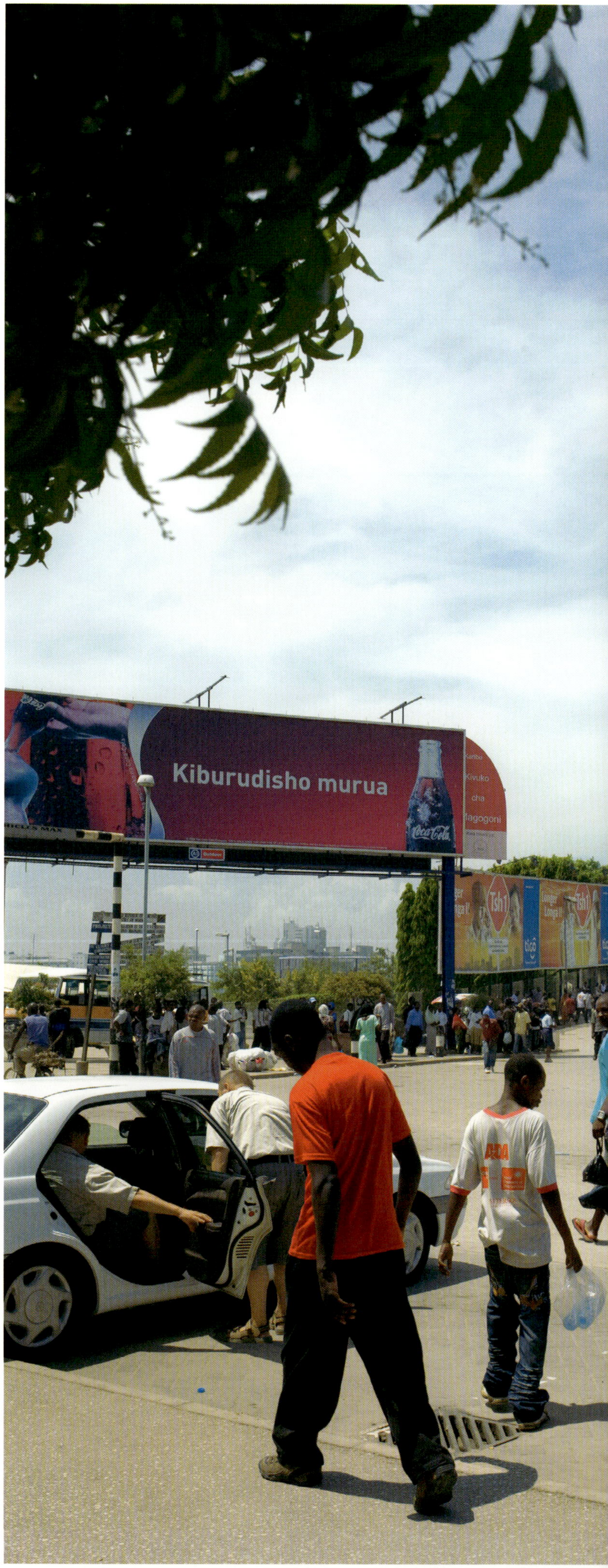

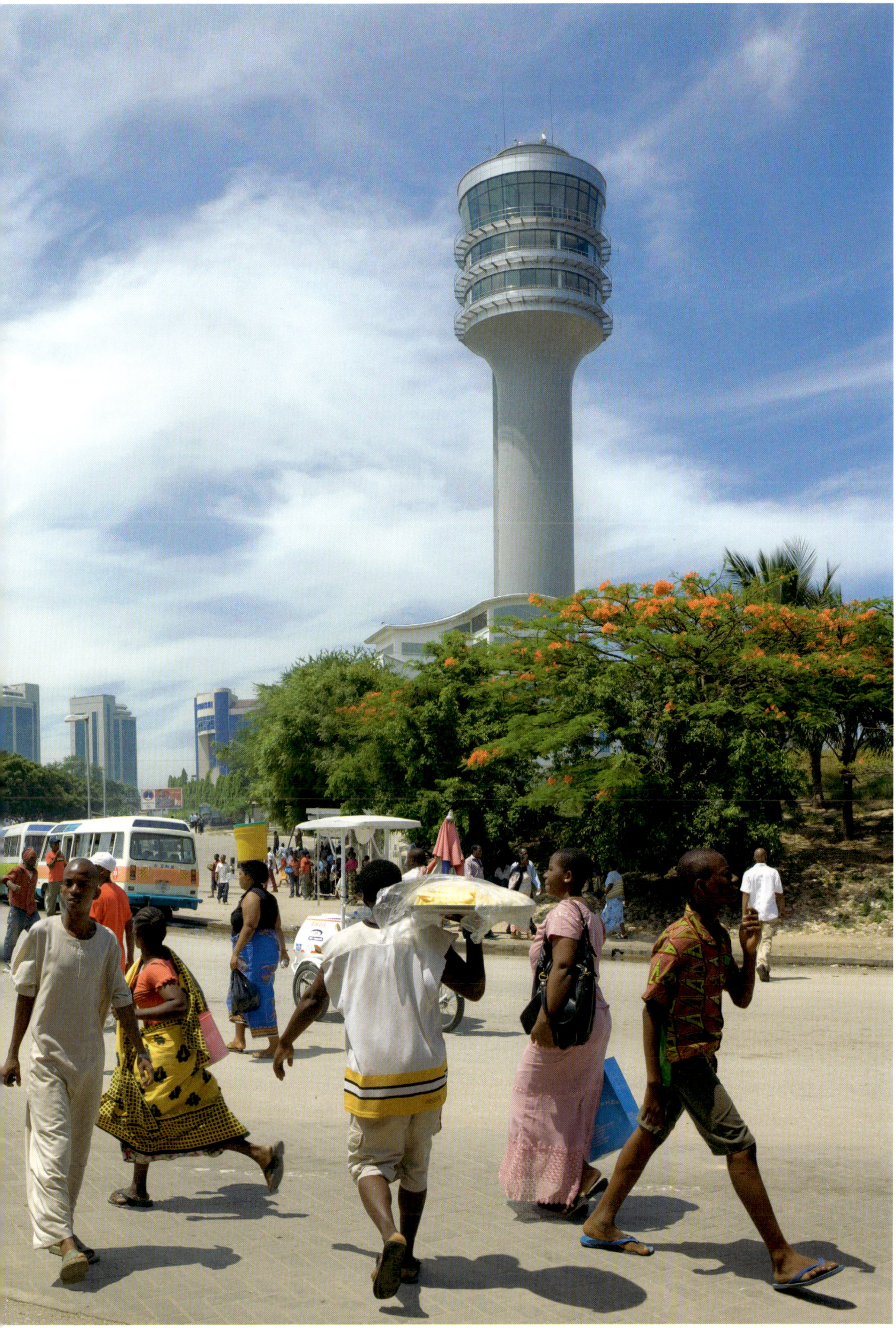

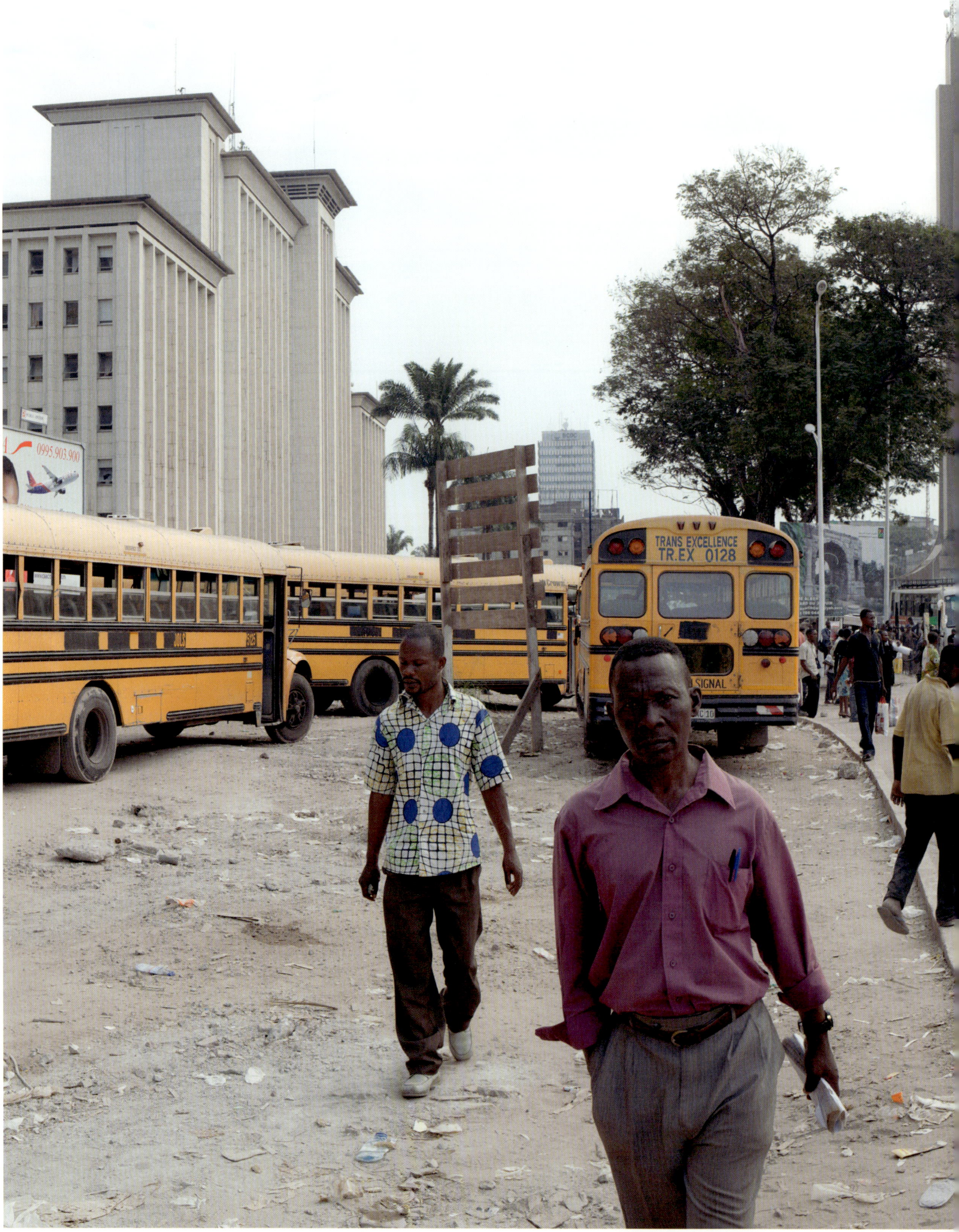

0995.903.900
TRANS EXCELLENCE
TR.EX 0128
SIGNAL

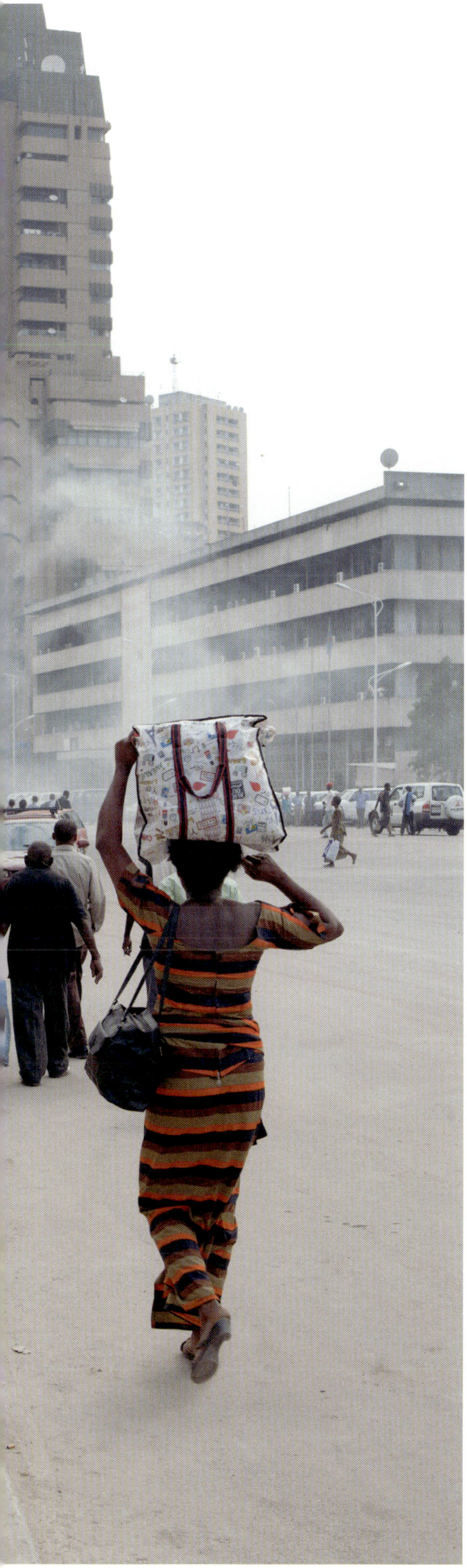

BOULEVARD 30 JUIN
KINSHASA

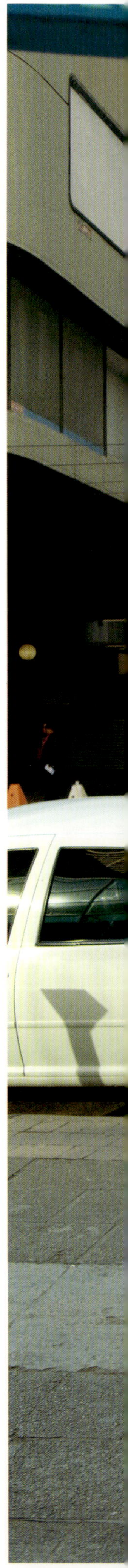

COMMISSIONER STREET
JOHANNESBURG

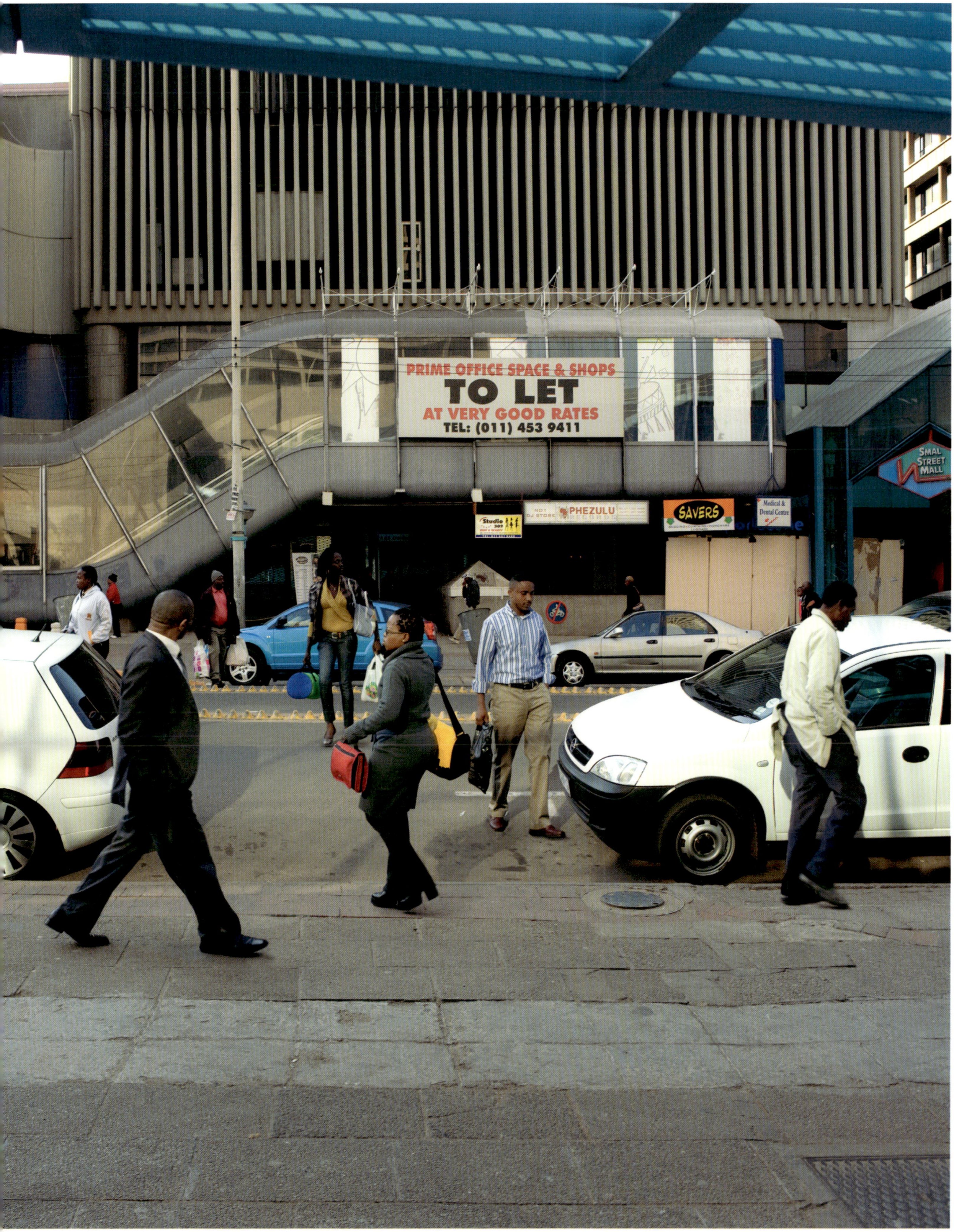
PRIME OFFICE SPACE & SHOPS
TO LET
AT VERY GOOD RATES
TEL: (011) 453 9411
PHEZULU
SAVERS
Medical & Dental Centre
SMAL STREET MALL

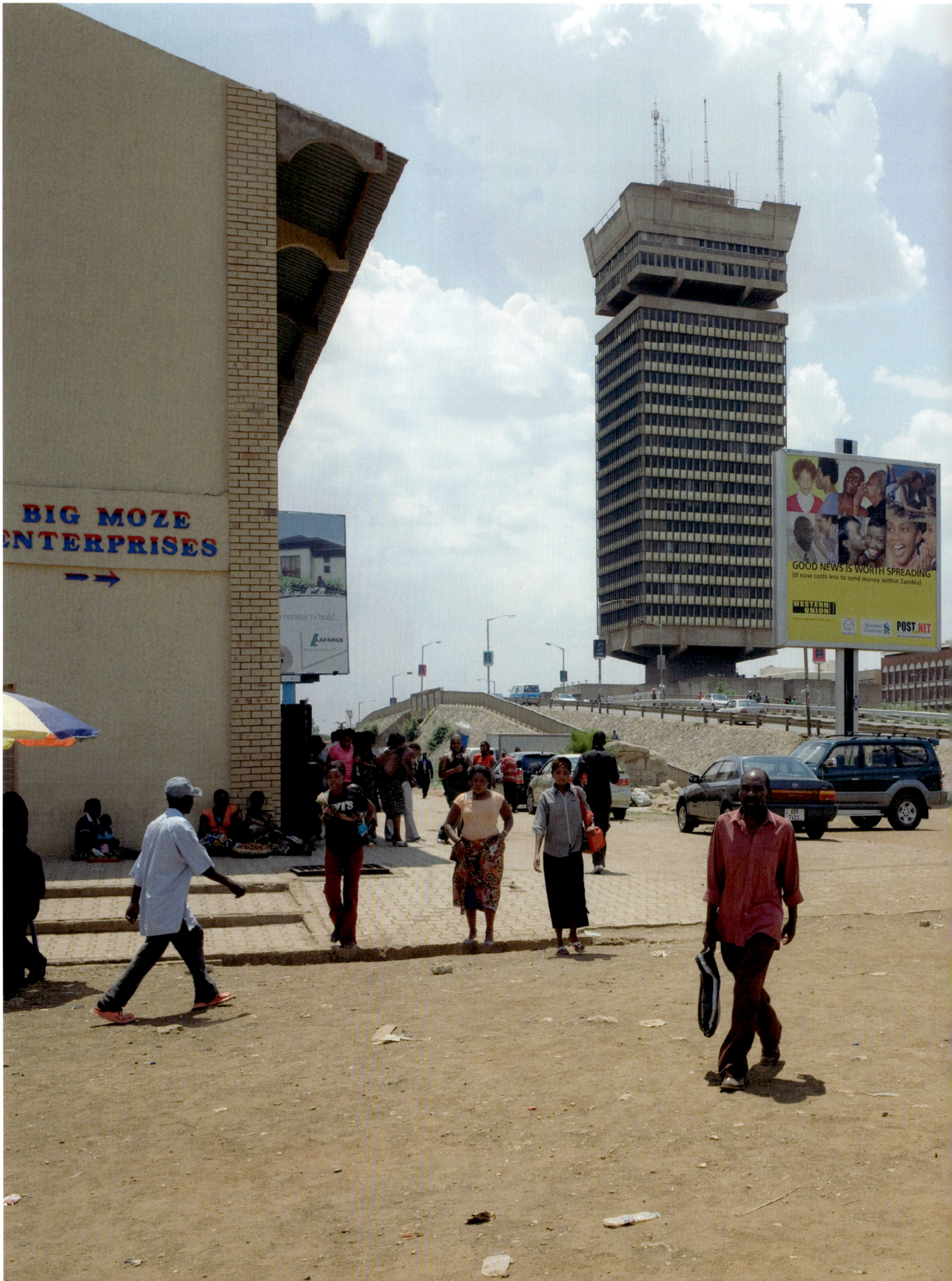
BIG MOZE
ENTERPRISES
GOOD NEWS IS WORTH SPREADING
POST NET

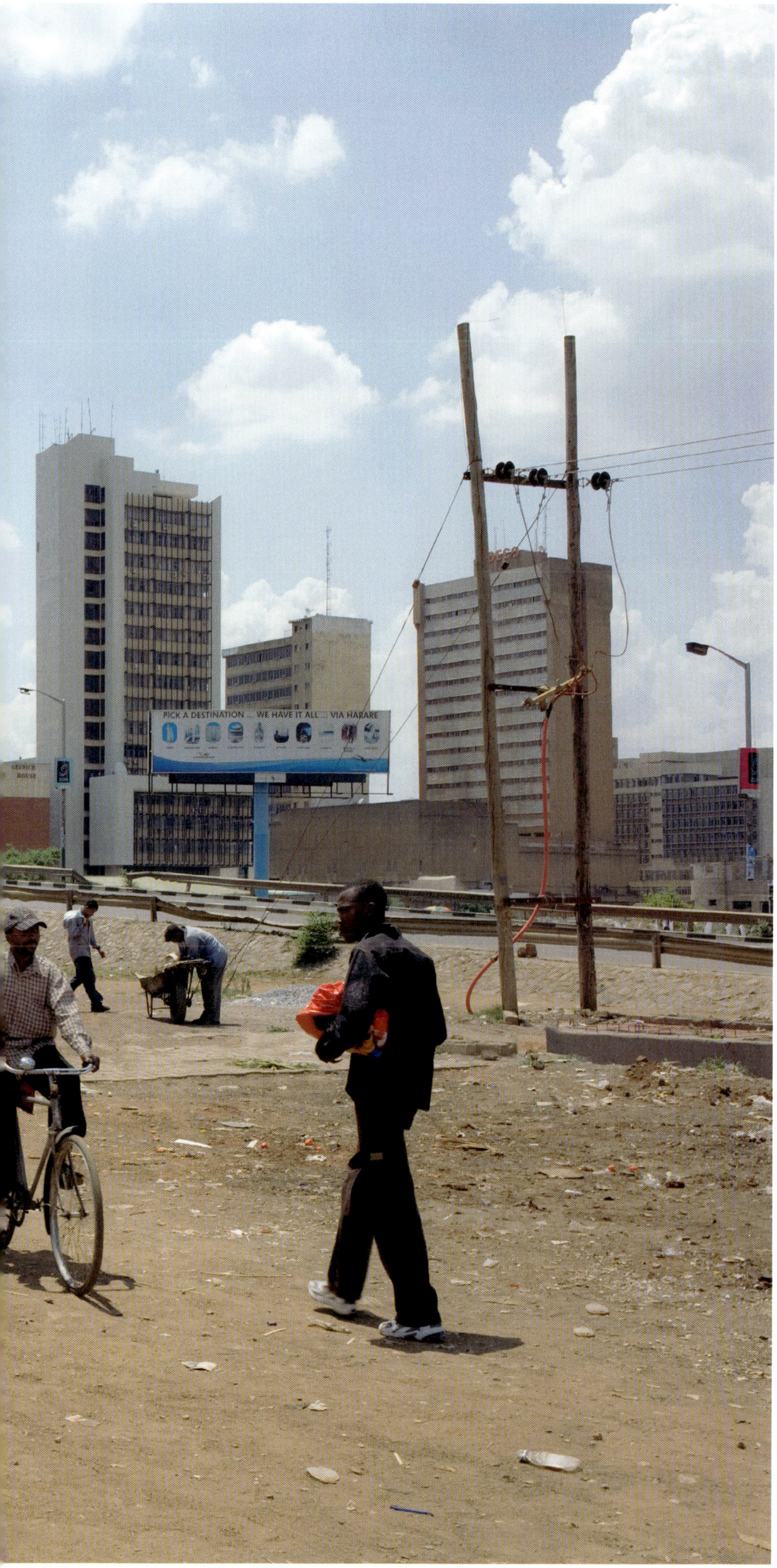

KAMWALA MARKET
LUSAKA

APAPA OWORONSOKI, OSHODI
LAGOS

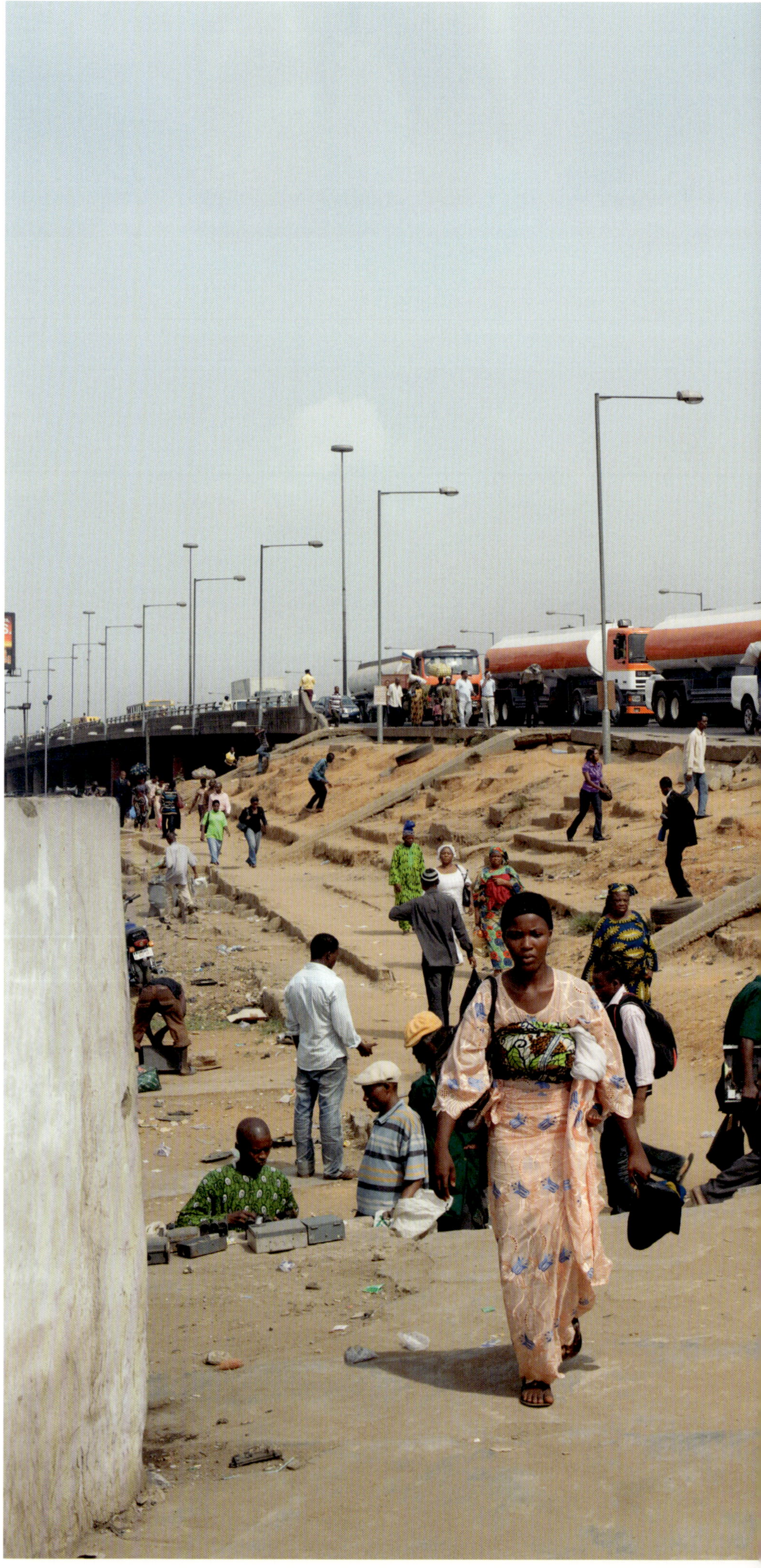

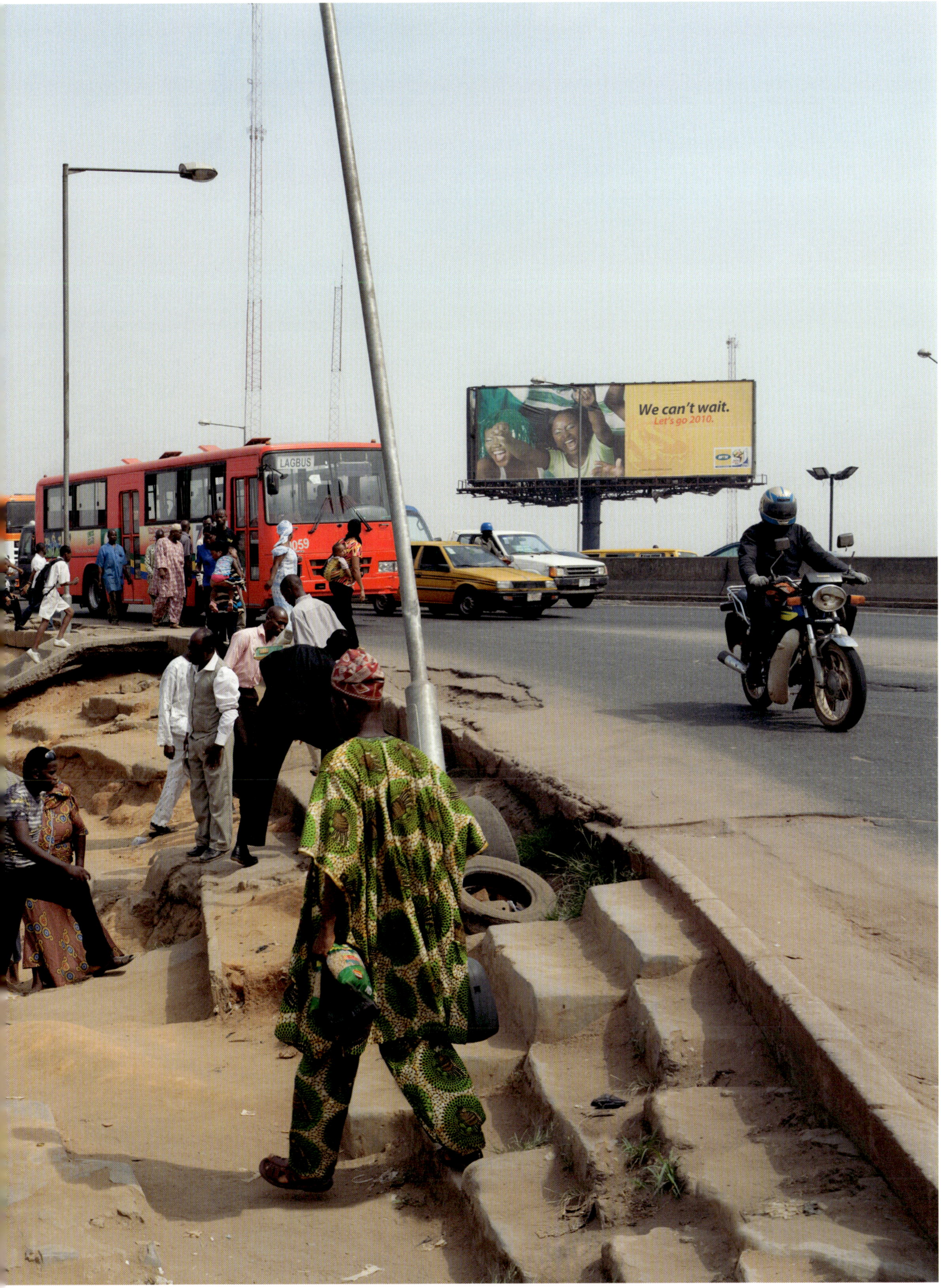
LAGBUS
We can't wait.
Let's go 2010.

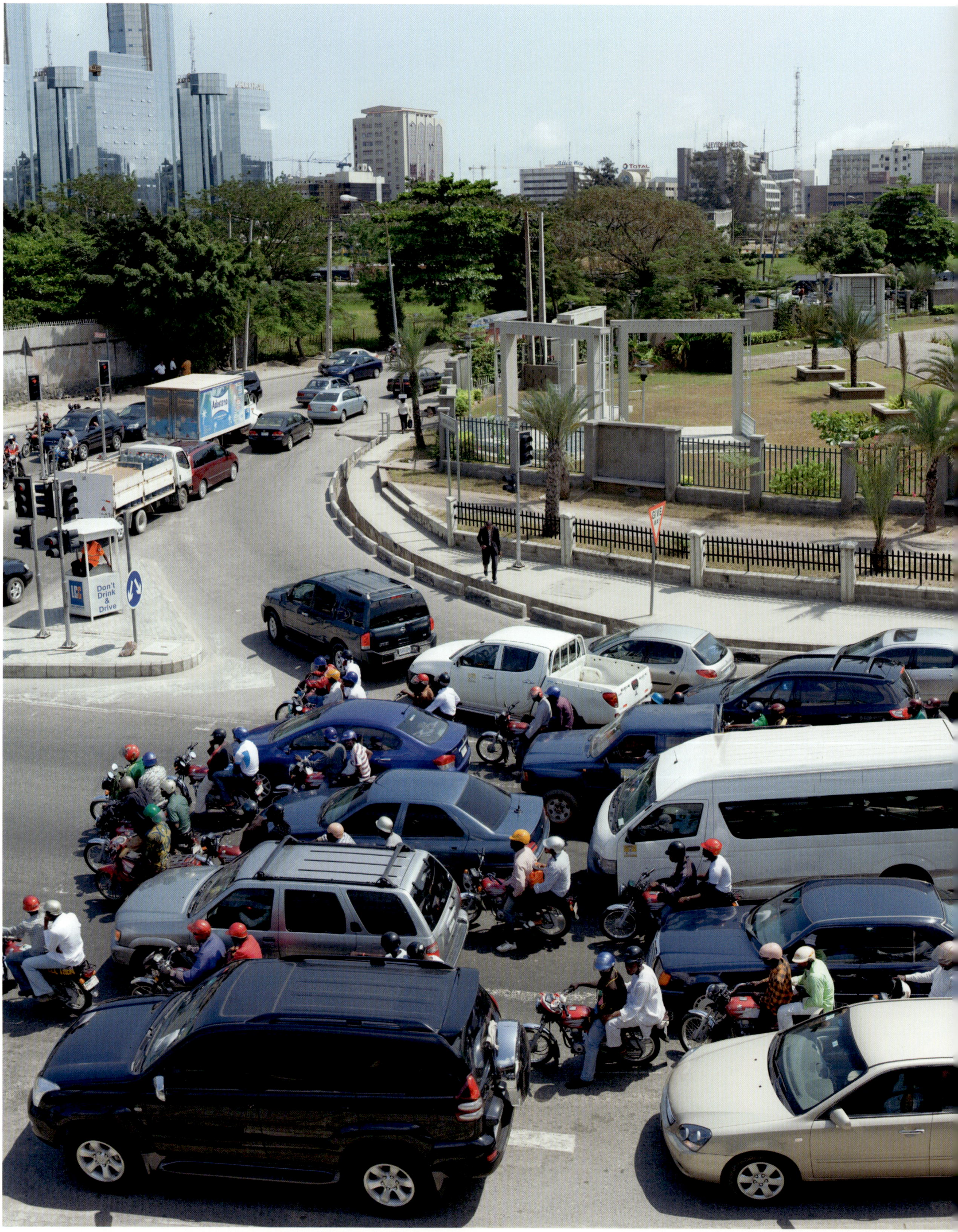
Don't
Drink
&
Drive

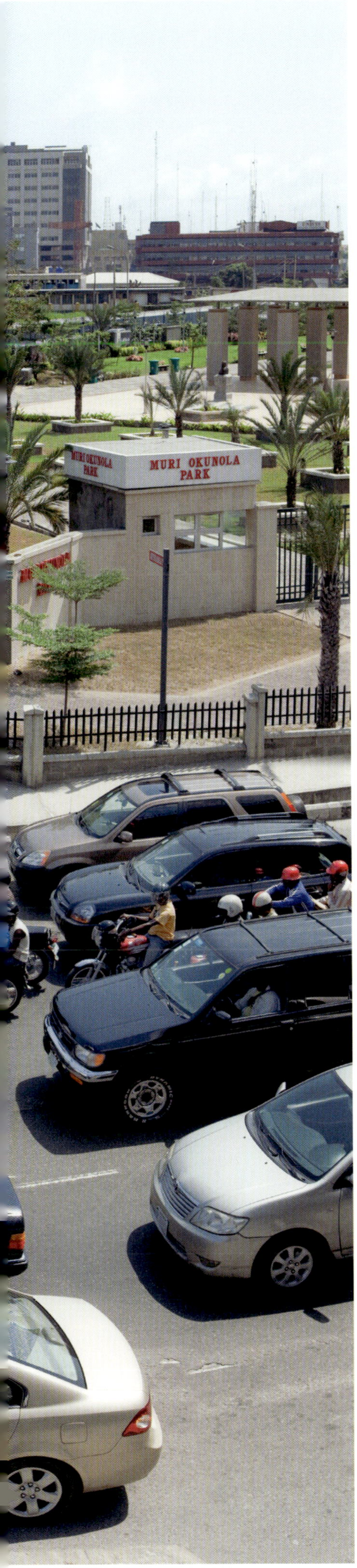

MAROKO ROAD, VICTORIA ISLAND
LAGOS

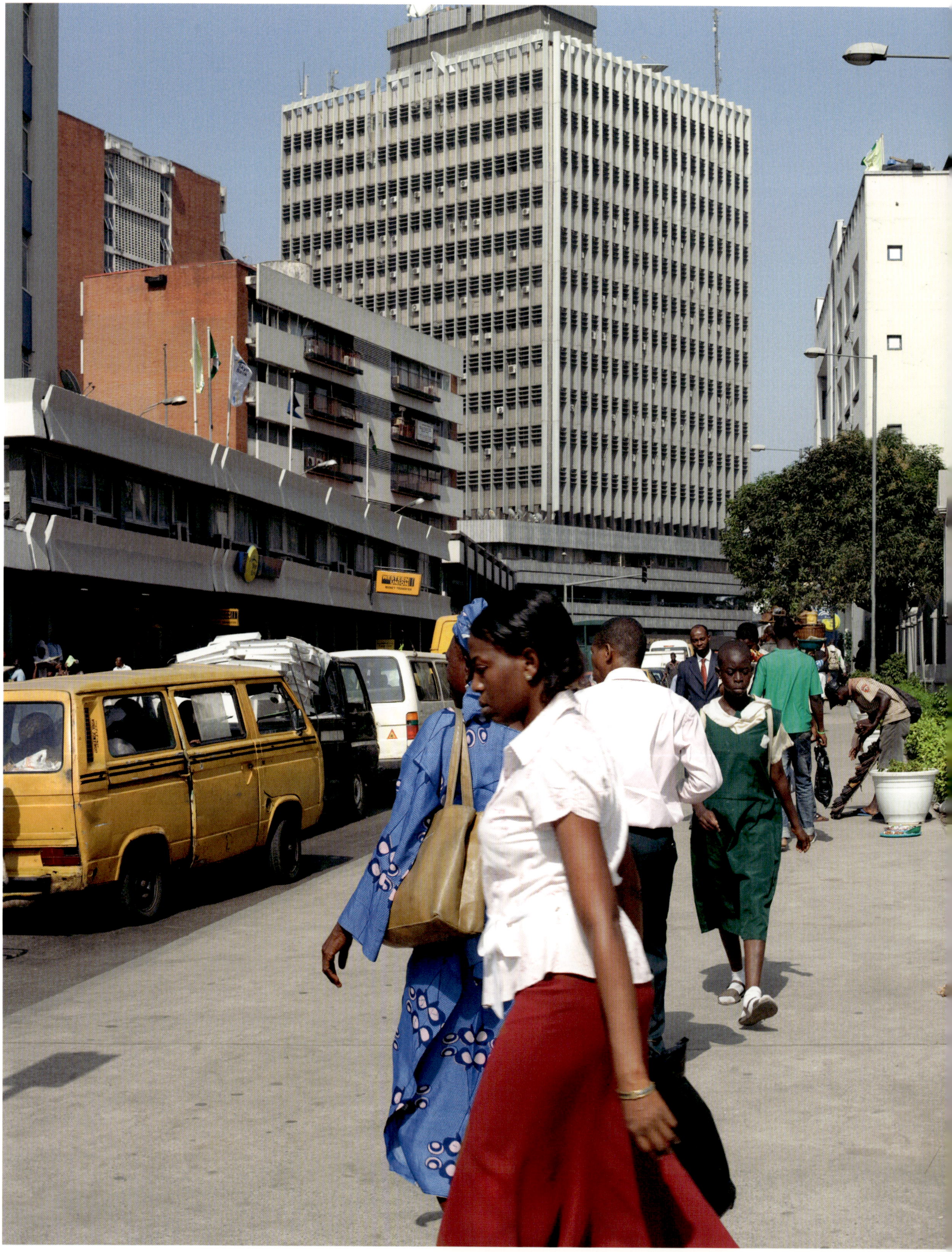
WESTERN UNION

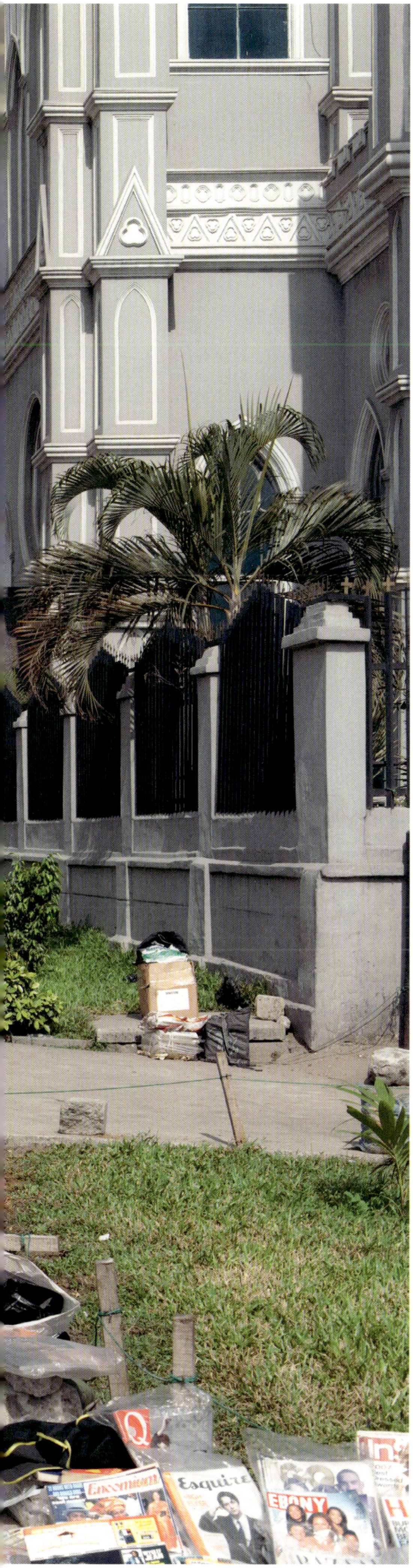
Q
Encomium
Esquire
EBONY

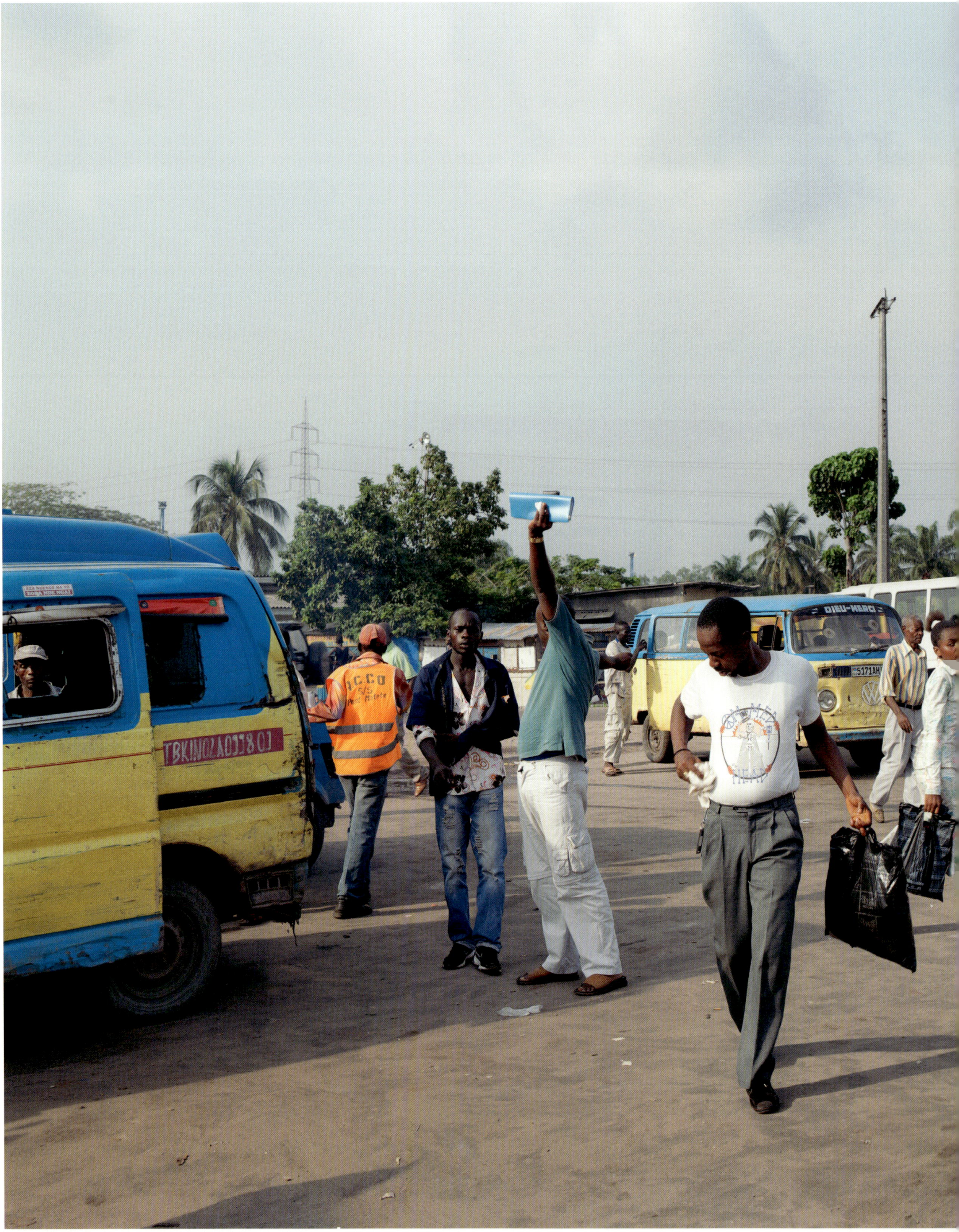
DIEU-MERCI

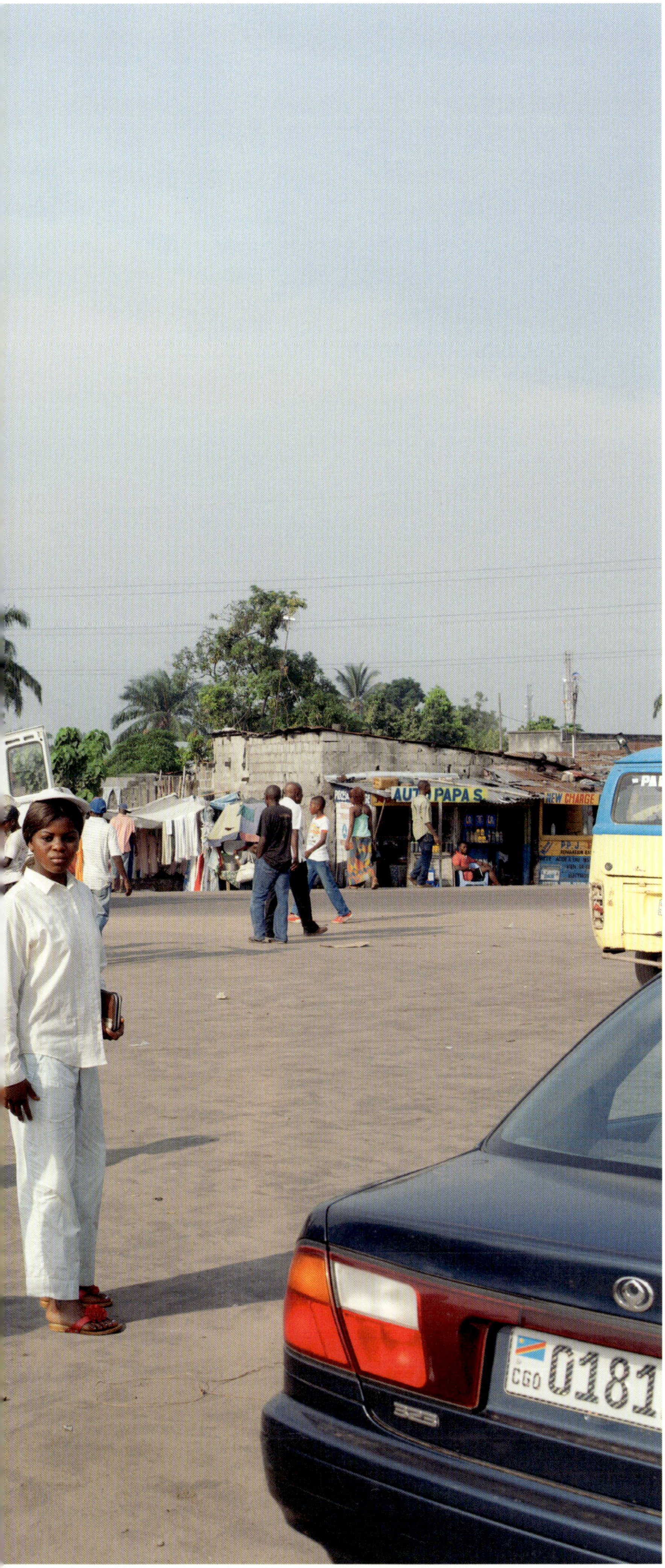

PONT MATETE TAXI
KINSHASA

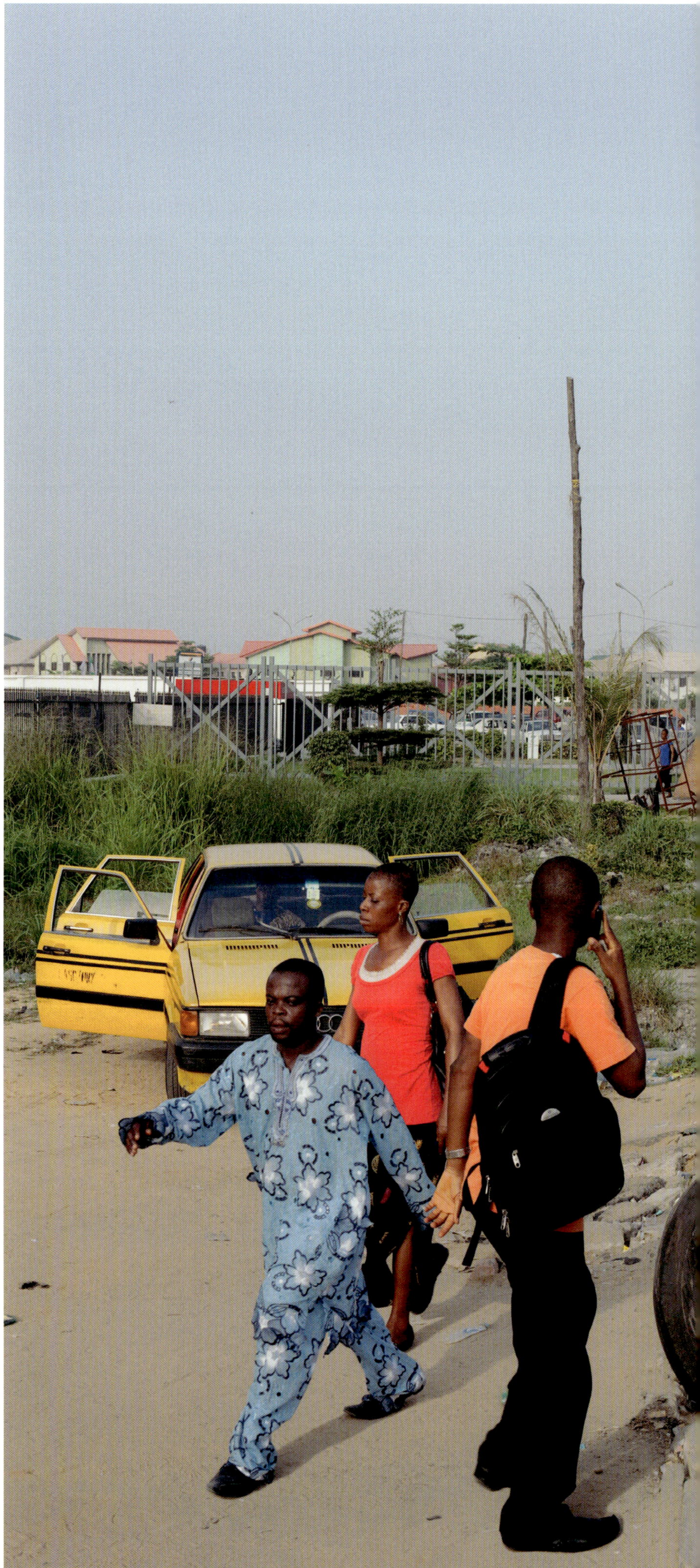

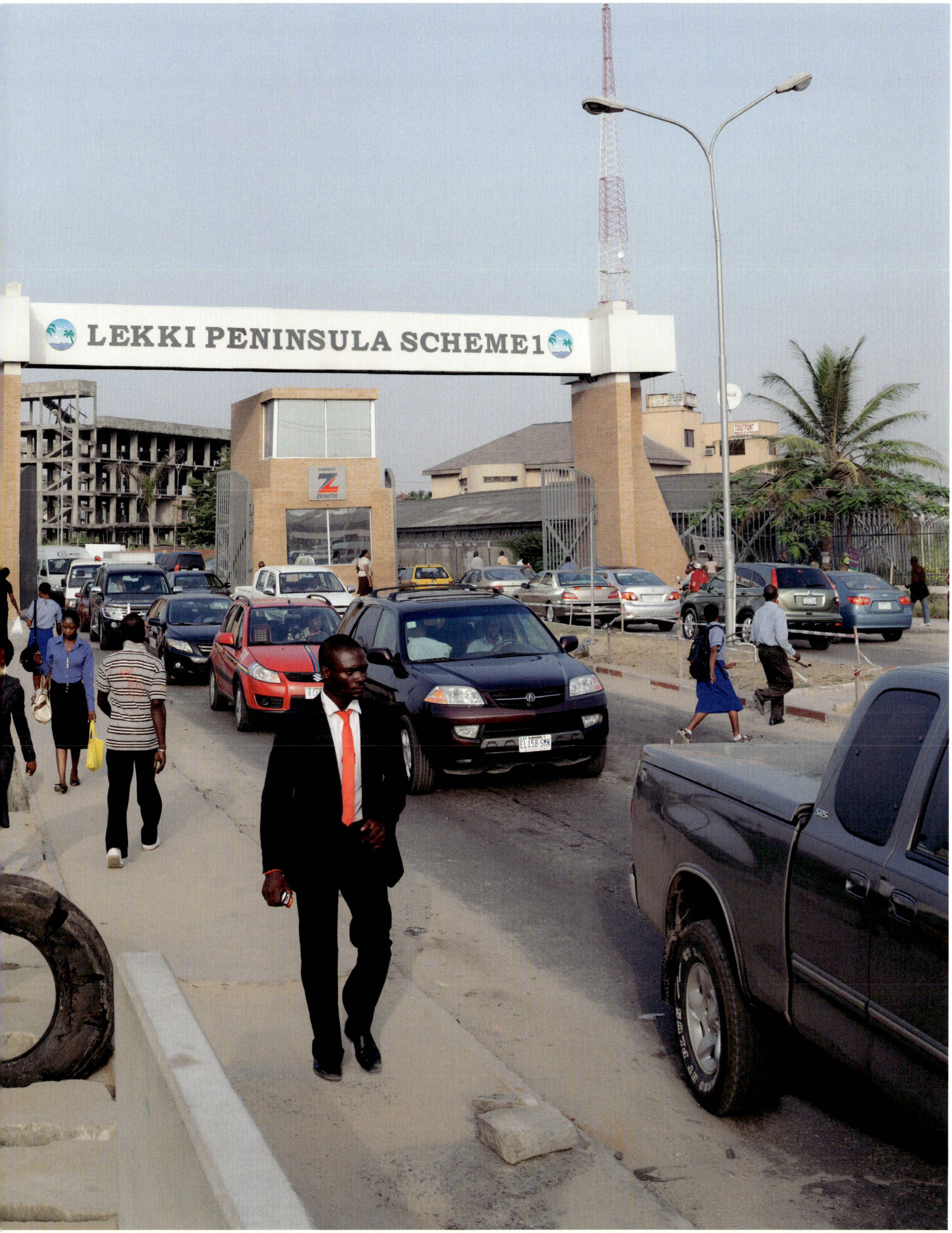
LEKKI PENINSULA SCHEME 1
ZENITH

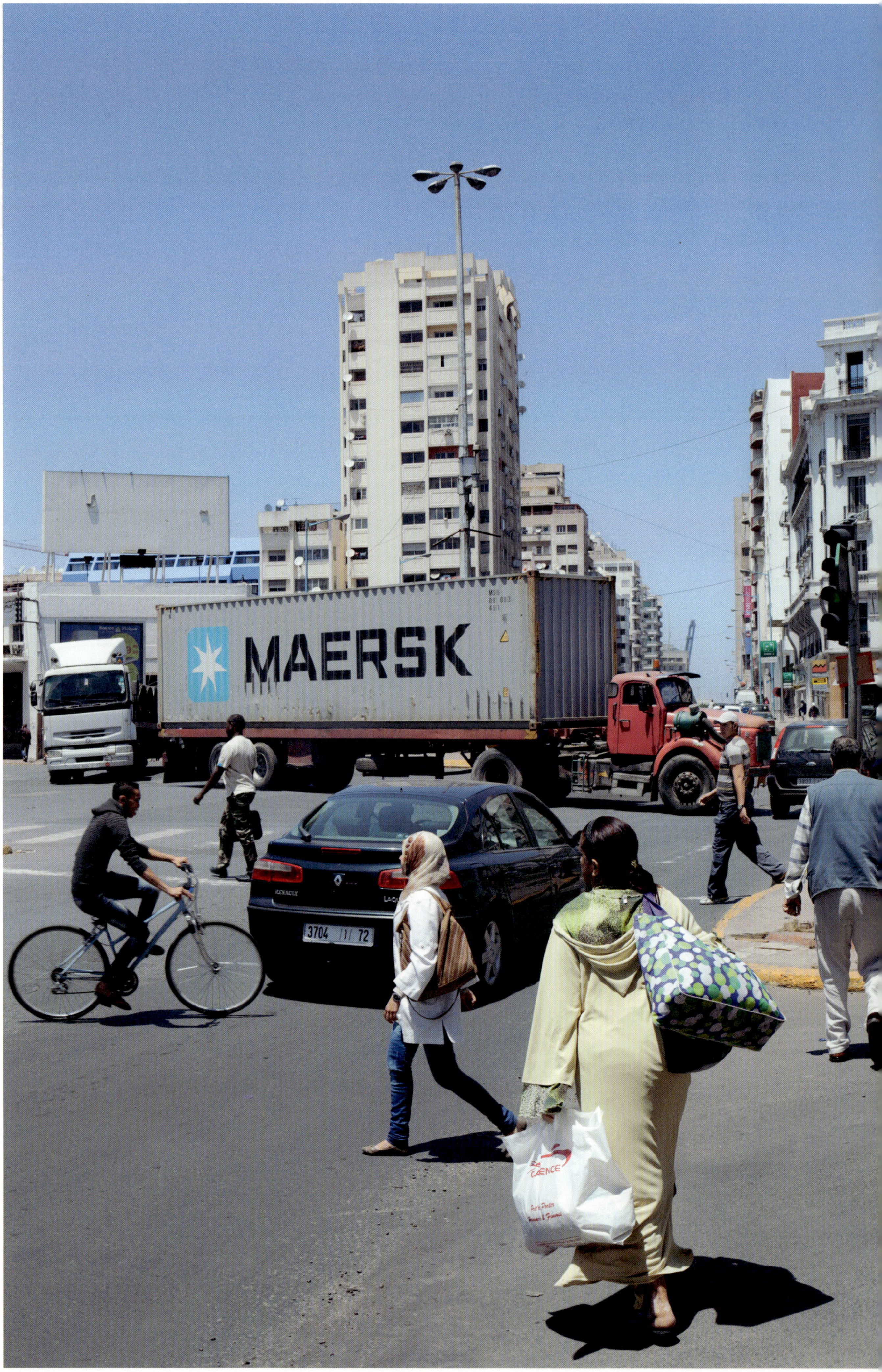
MAERSK

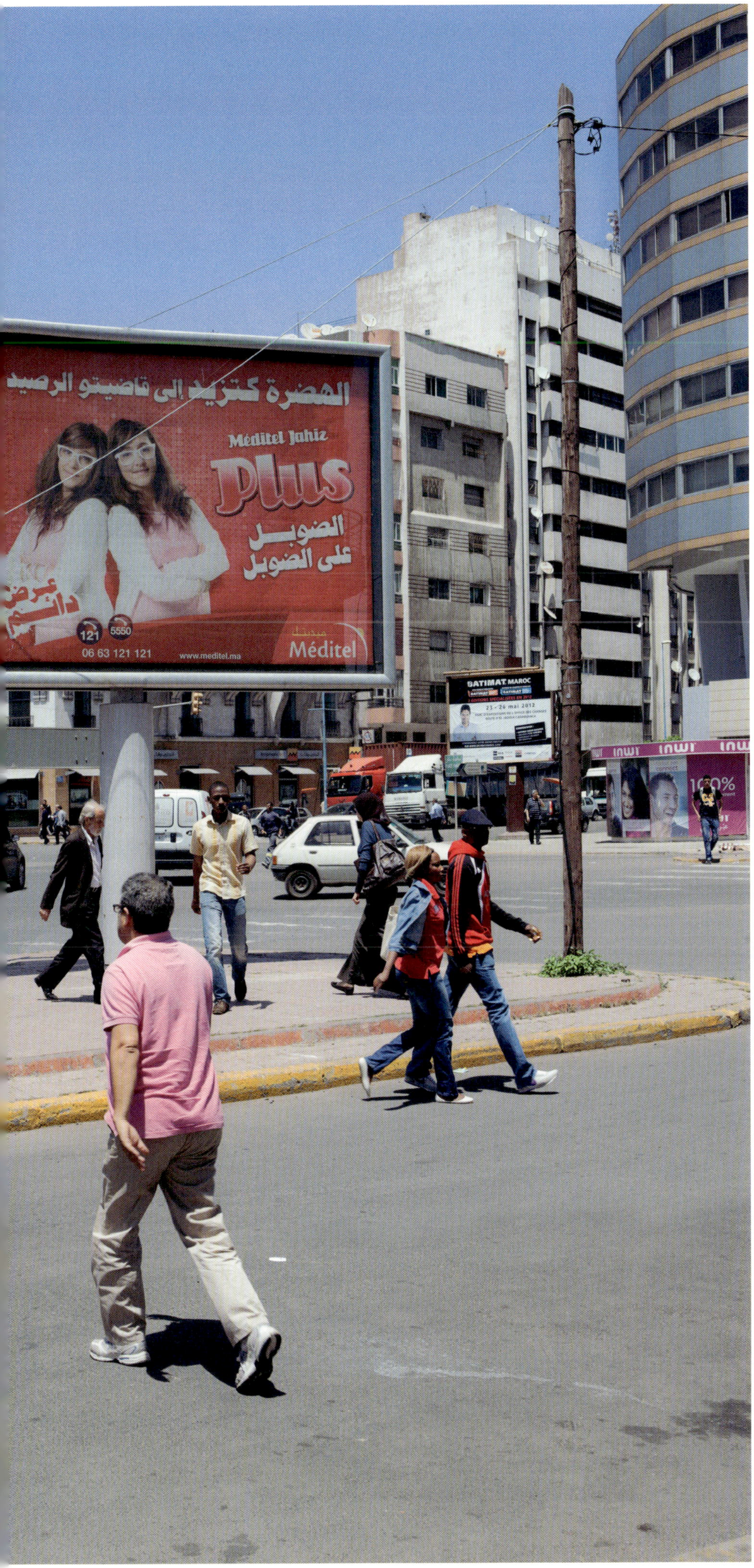

PLACE DAKAR
CASABLANCA

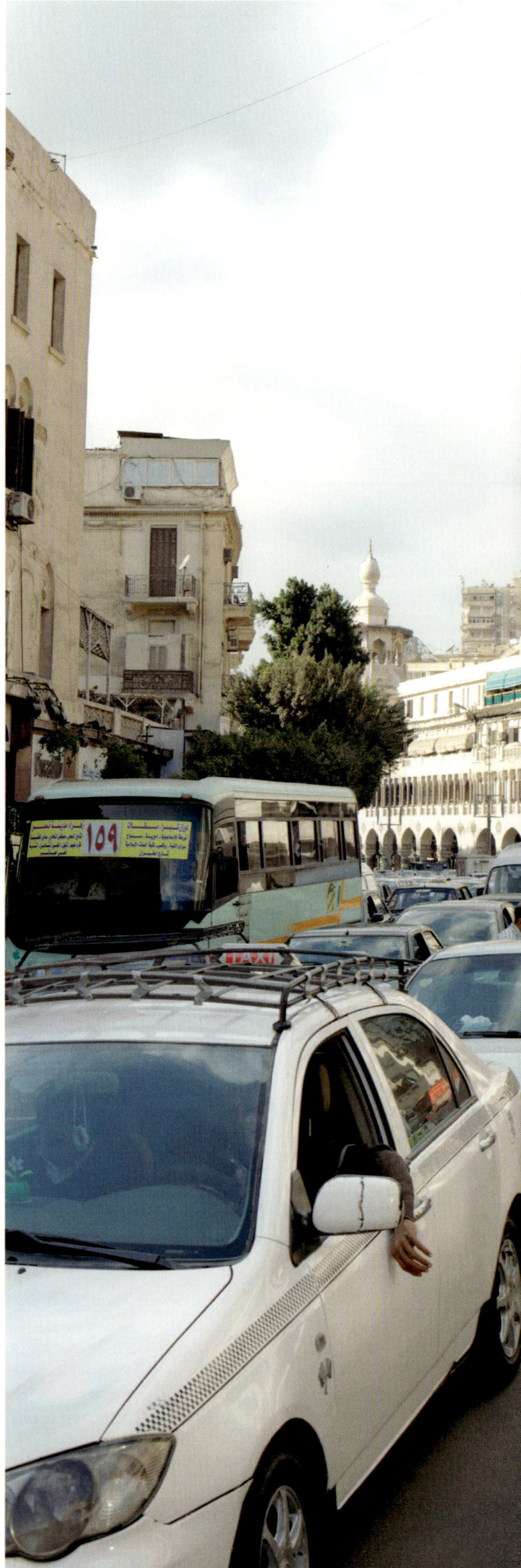

AL AHRAM STREET, HELIOPOLIS
CAIRO

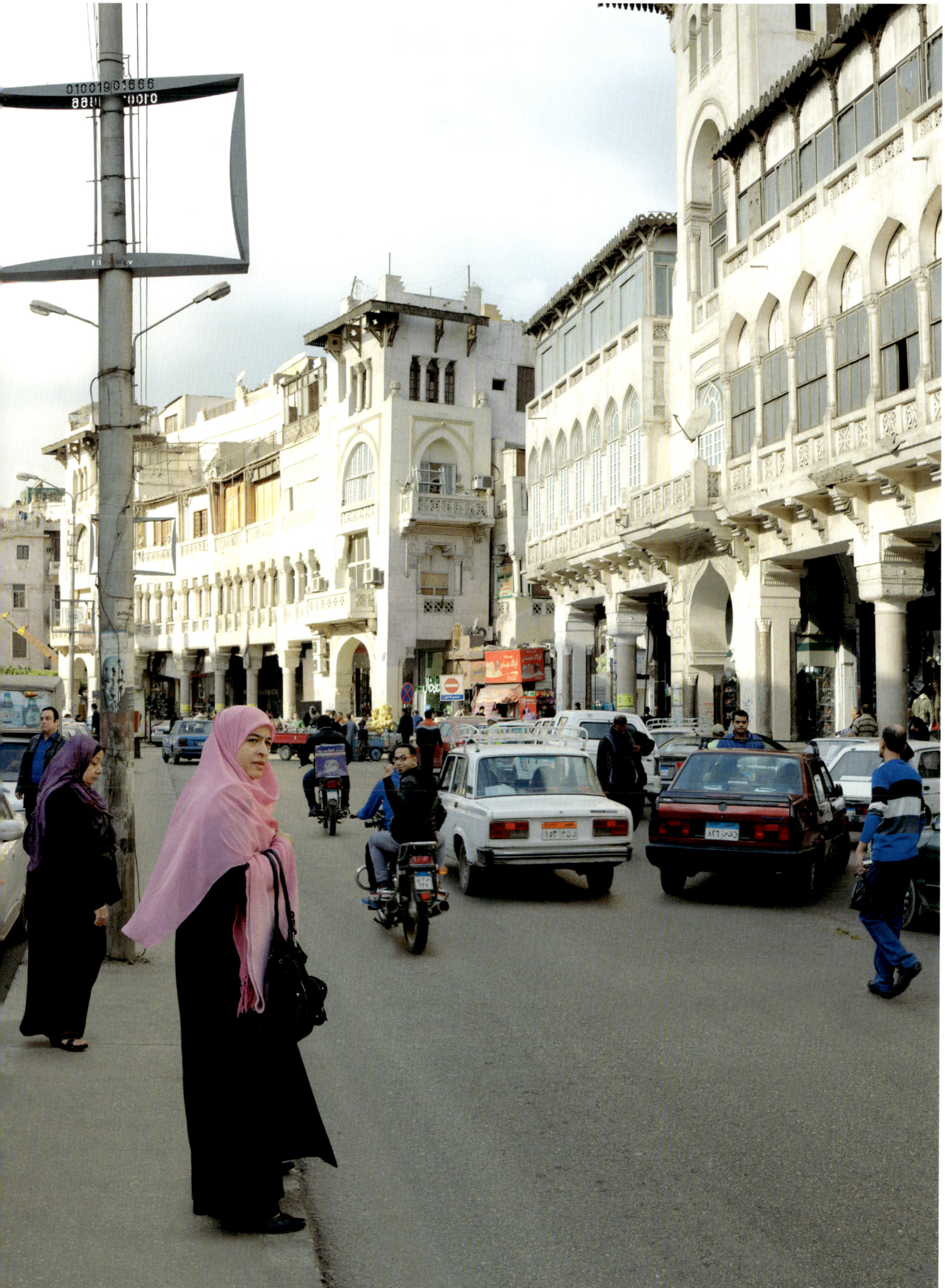
01001901666

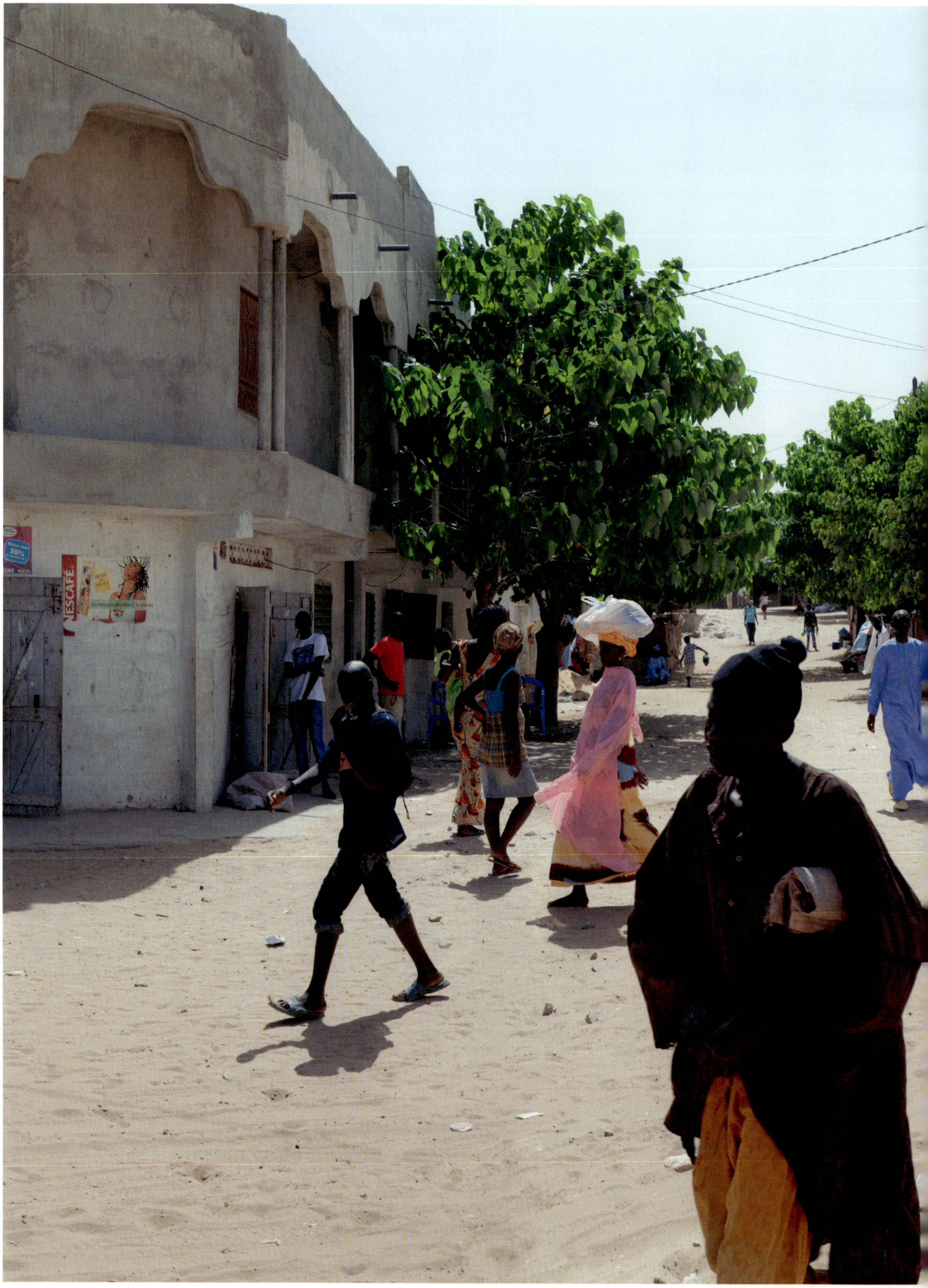

MAIRIE, PIKINE
DAKAR

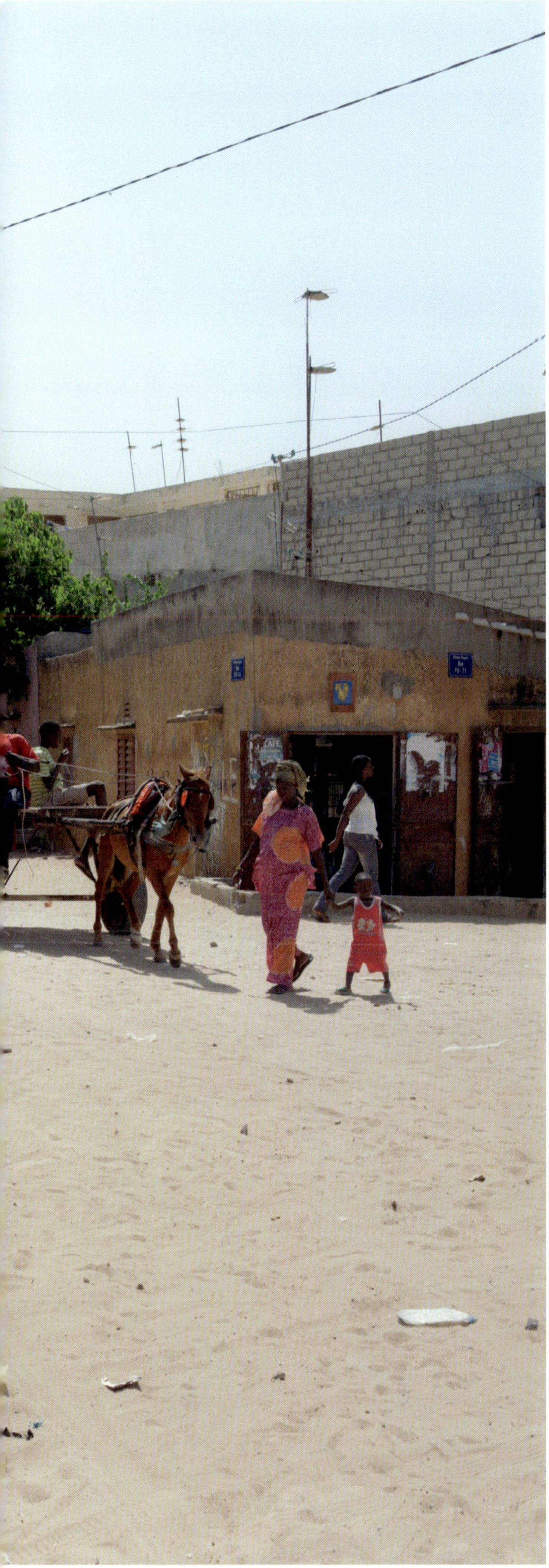

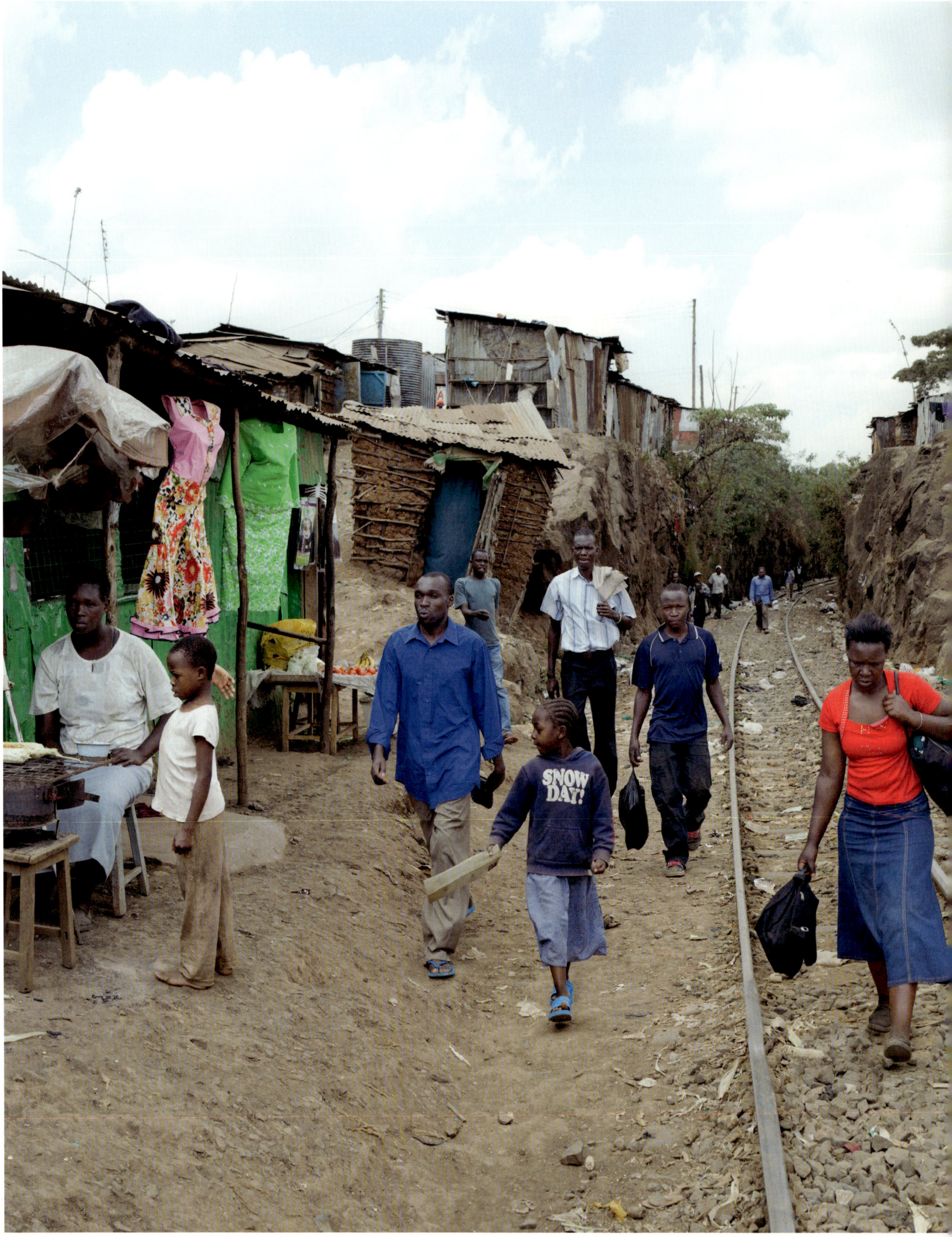
SNOW
DAY!

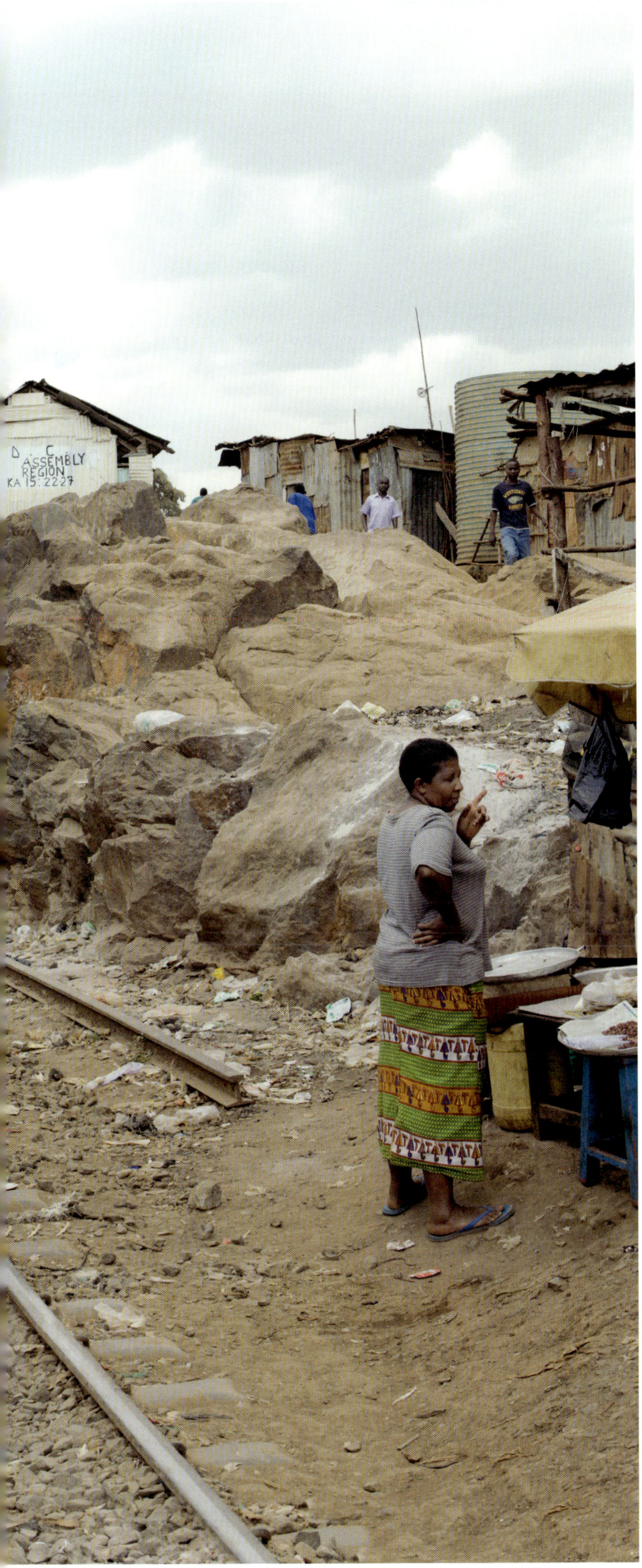

LINDI, KIBERA
NAIROBI

MULUNGUSHI ROAD
LUSAKA

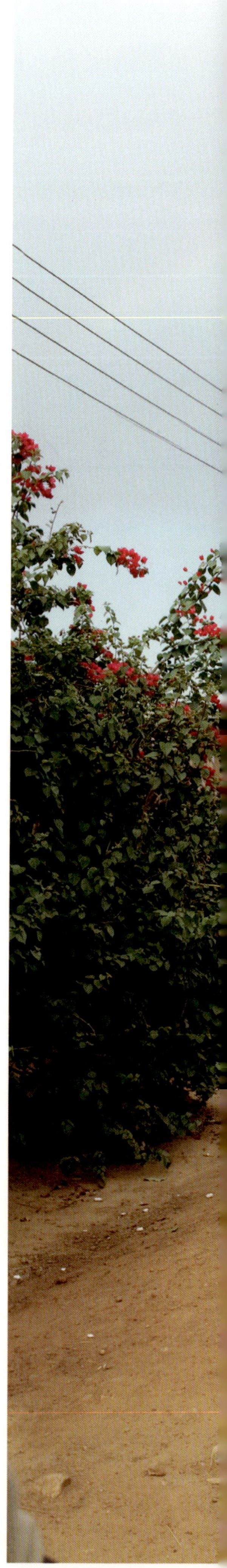

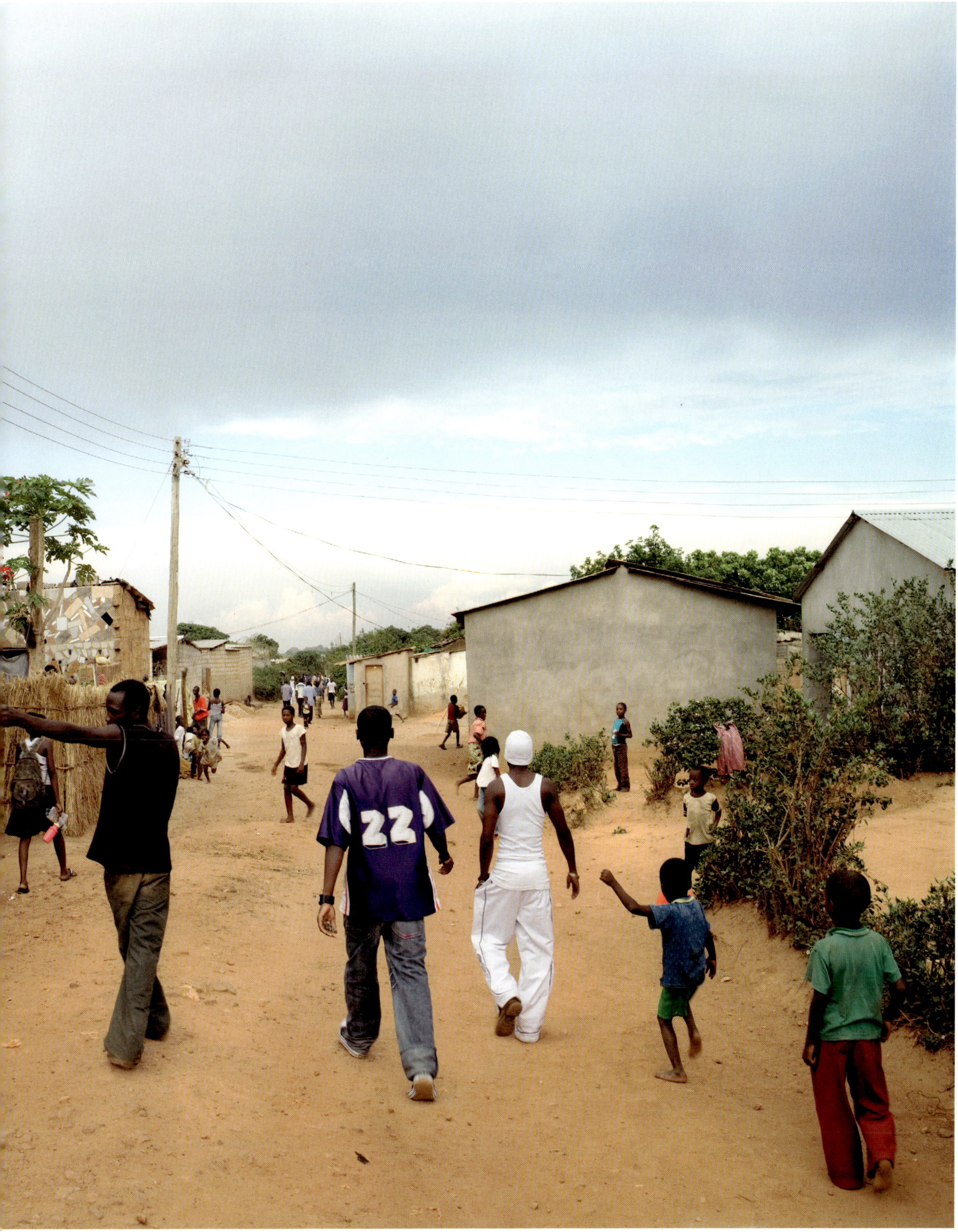
22

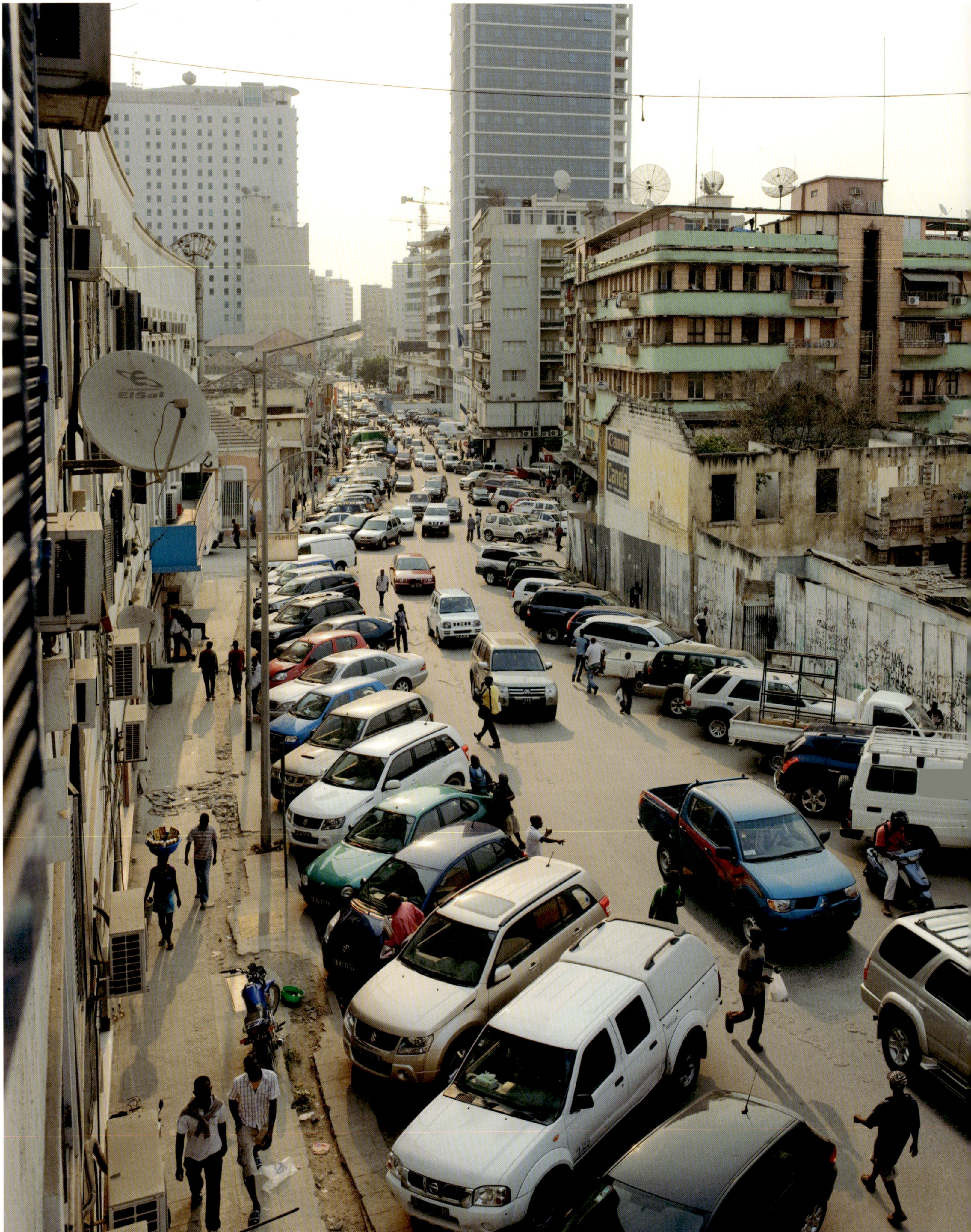
ElSat
Canon

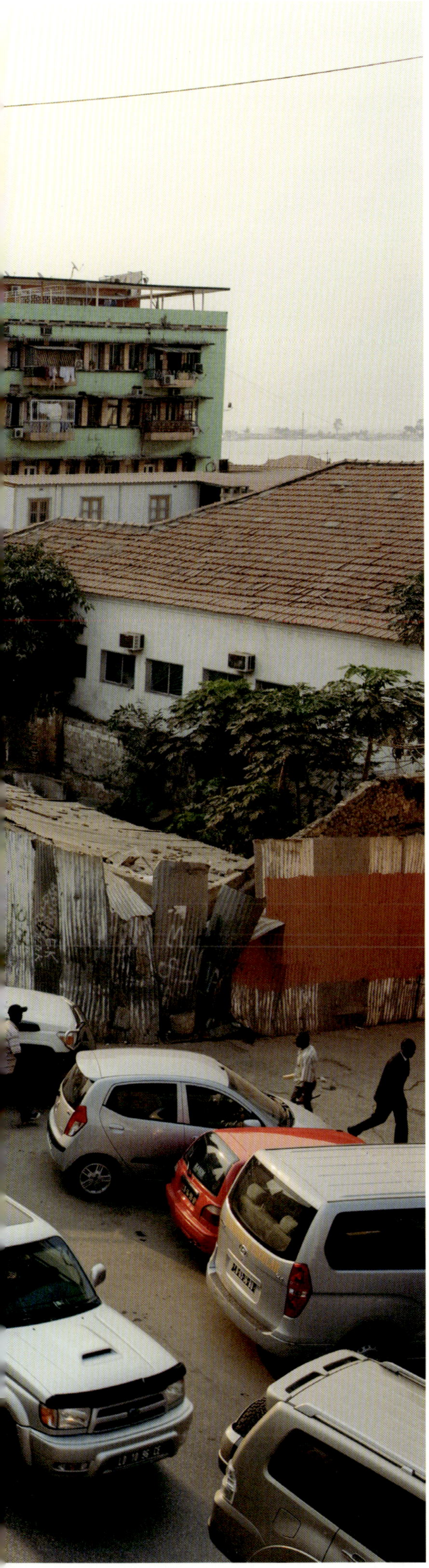

RUA RAINHA GINGA
LUANDA

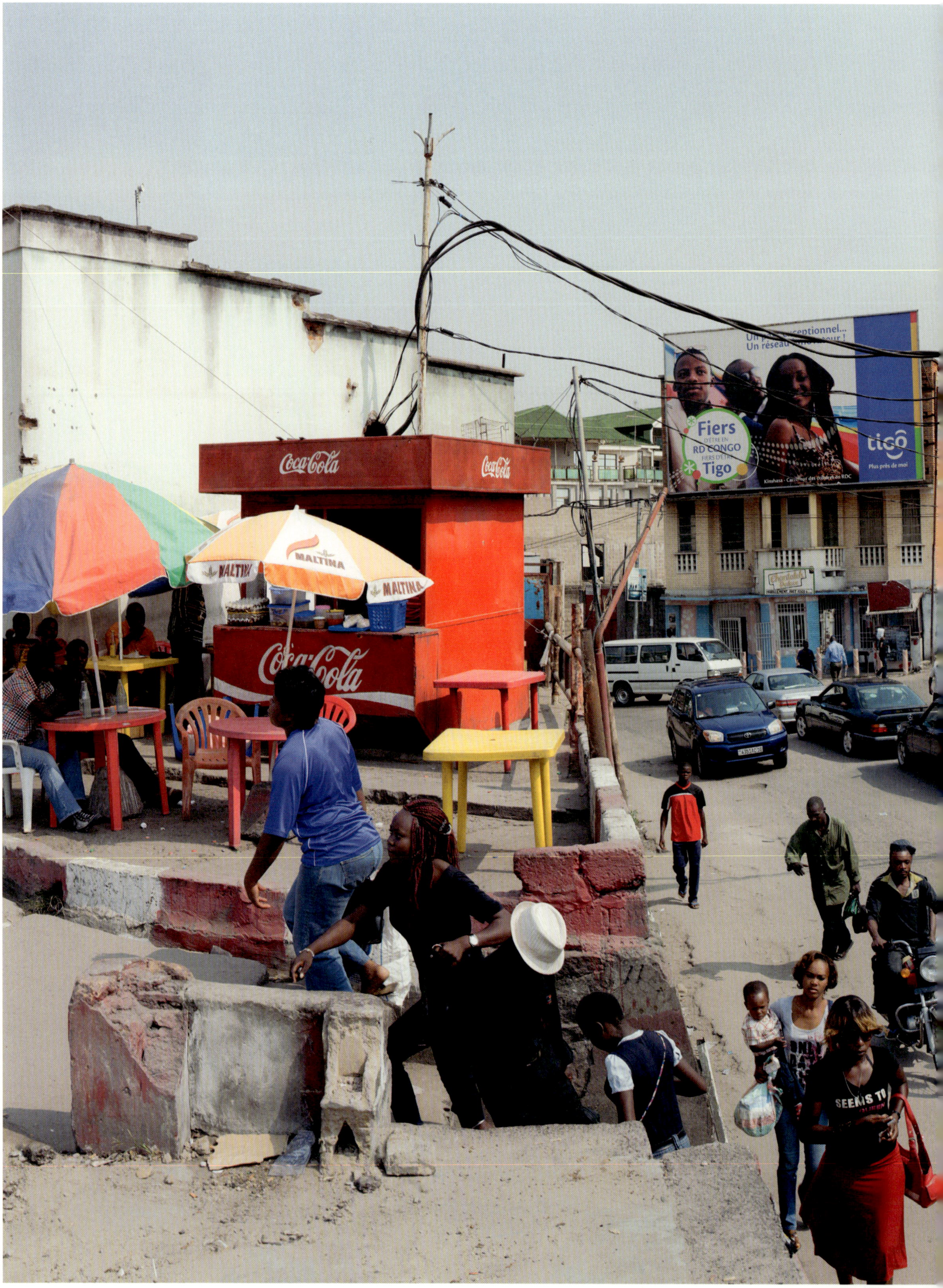
Coca-Cola
Coca-Cola
MALTINA
MALTINA
Coca-Cola
Fiers
RD CONGO
Tigo
tigo
Plus près de moi
SEEMS TO

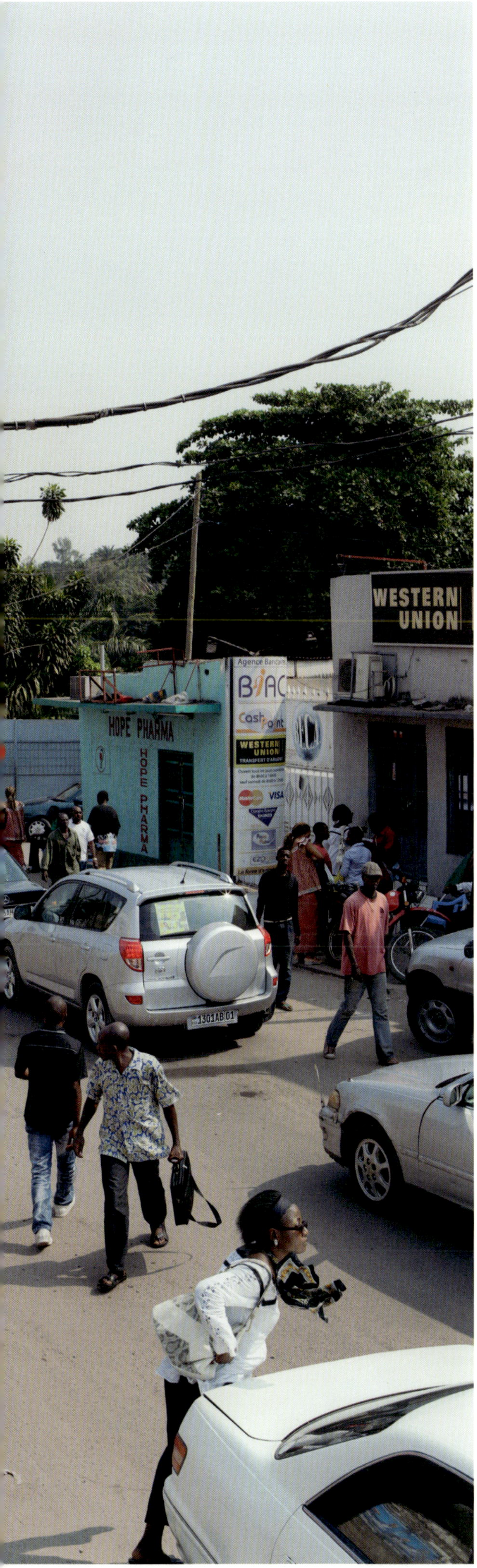

RUE NGUMA, MAGAZIN
KINSHASA

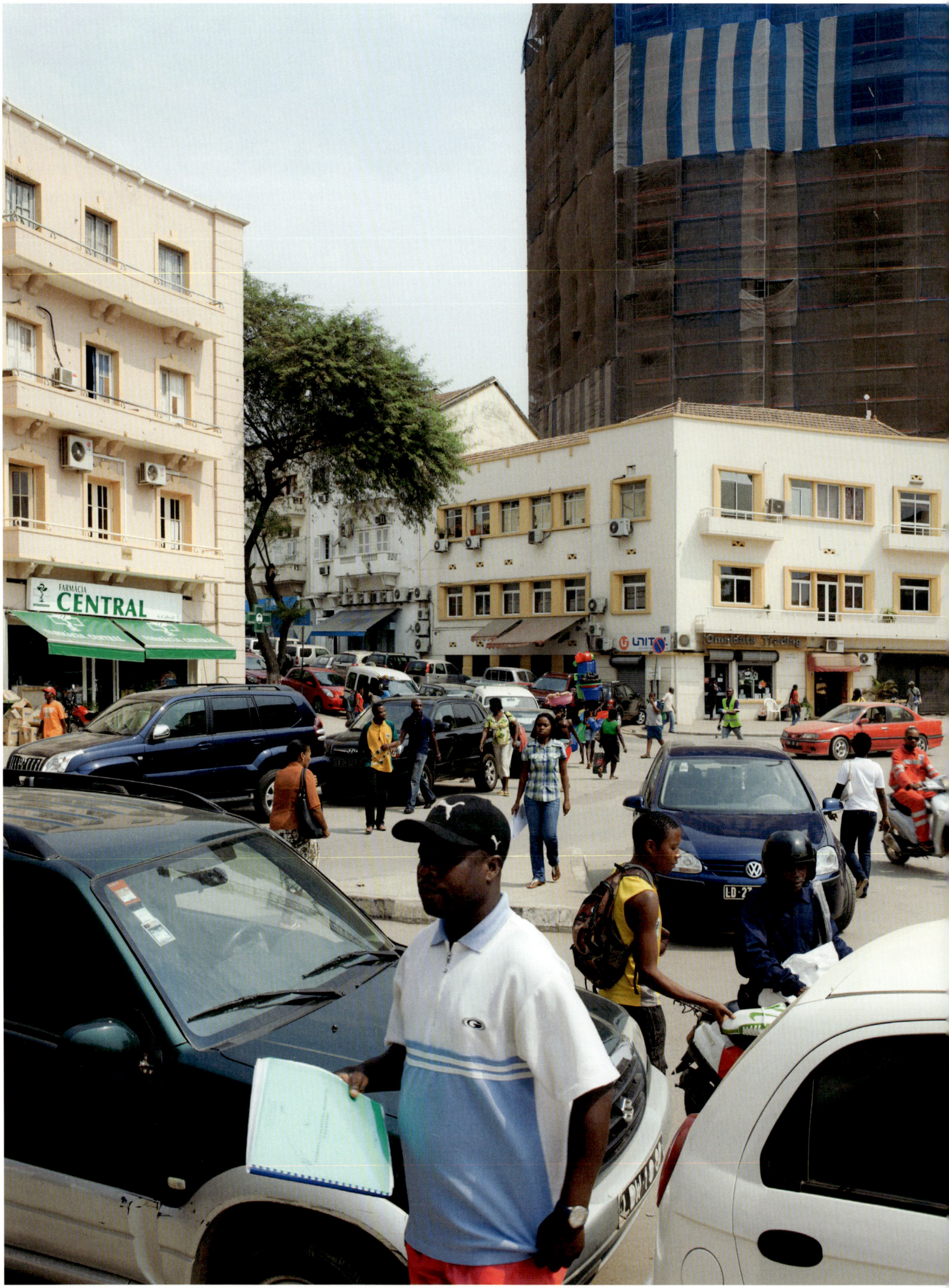
FARMÁCIA
CENTRAL

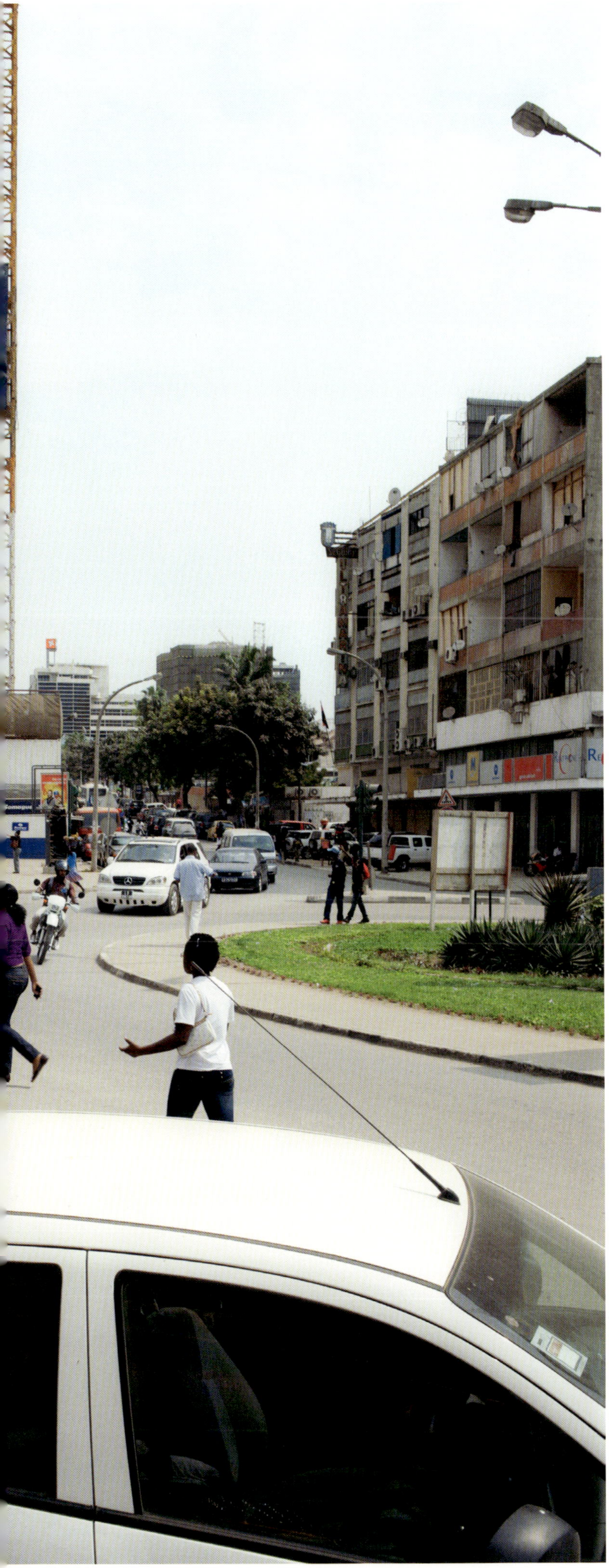

LARGO AMILCAR CABRAL
LUANDA

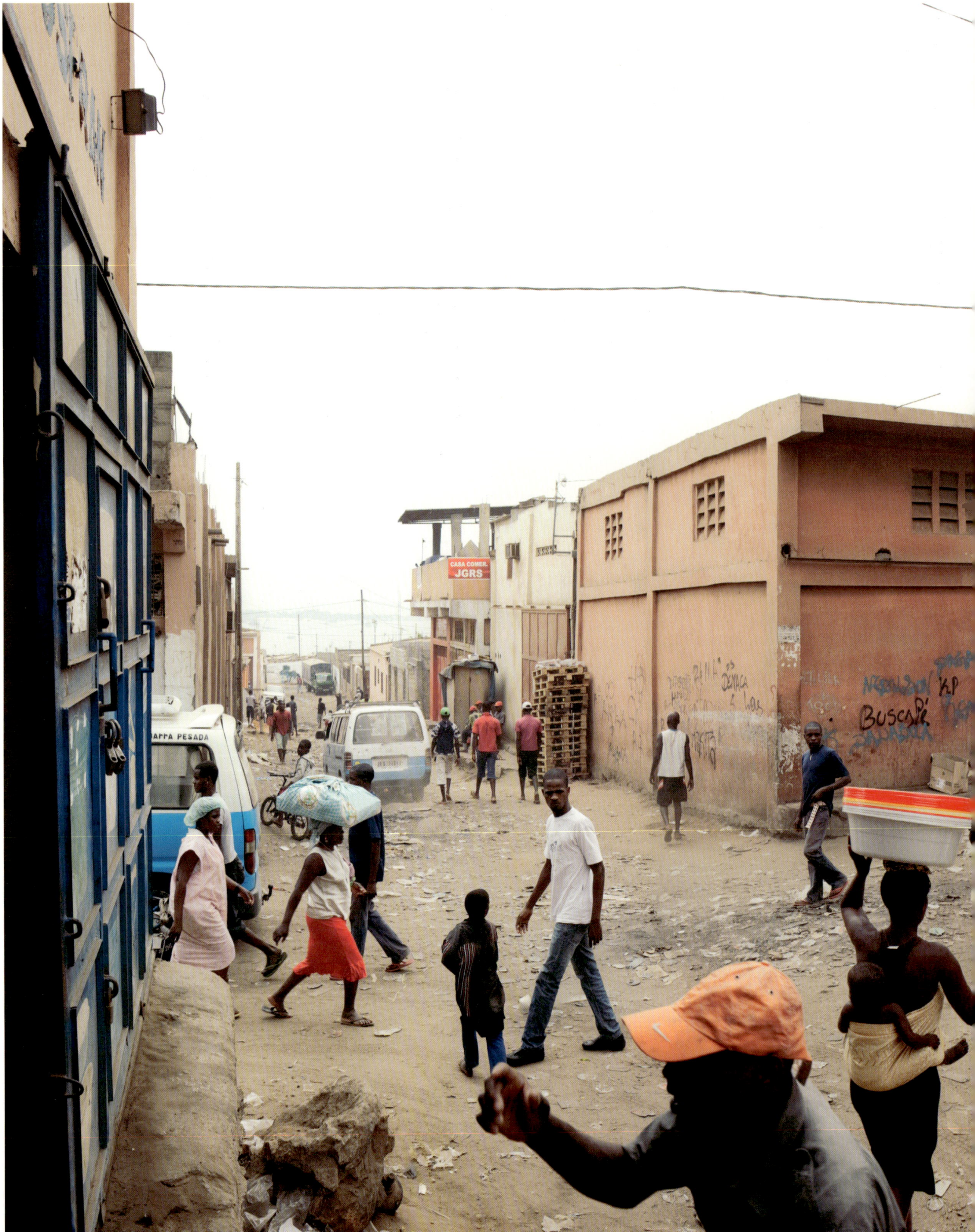
CASA COMER.
JGRS

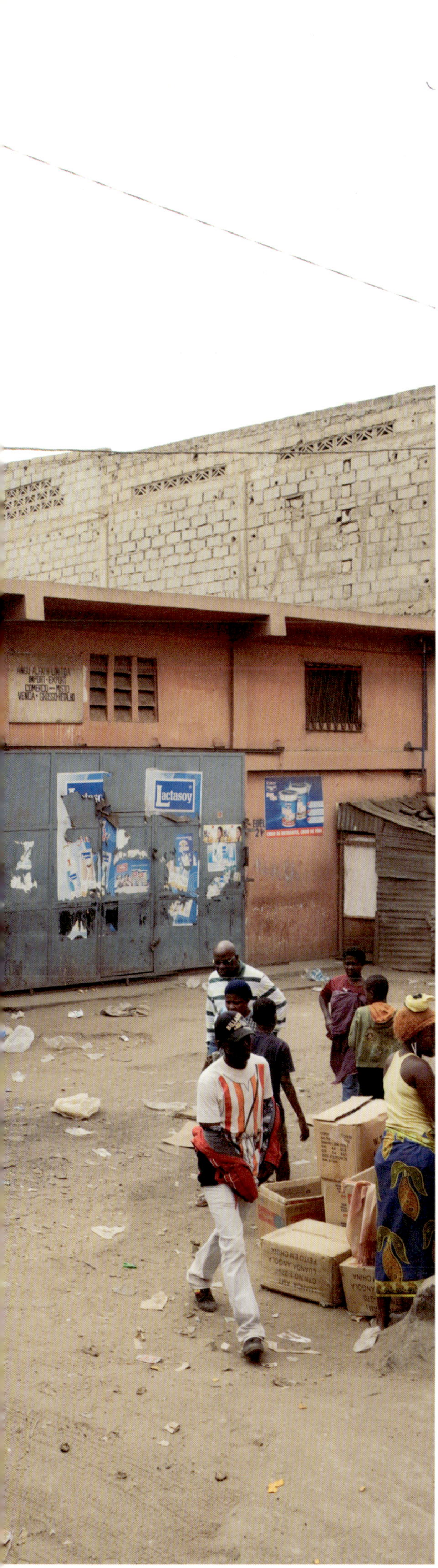
Lactasoy

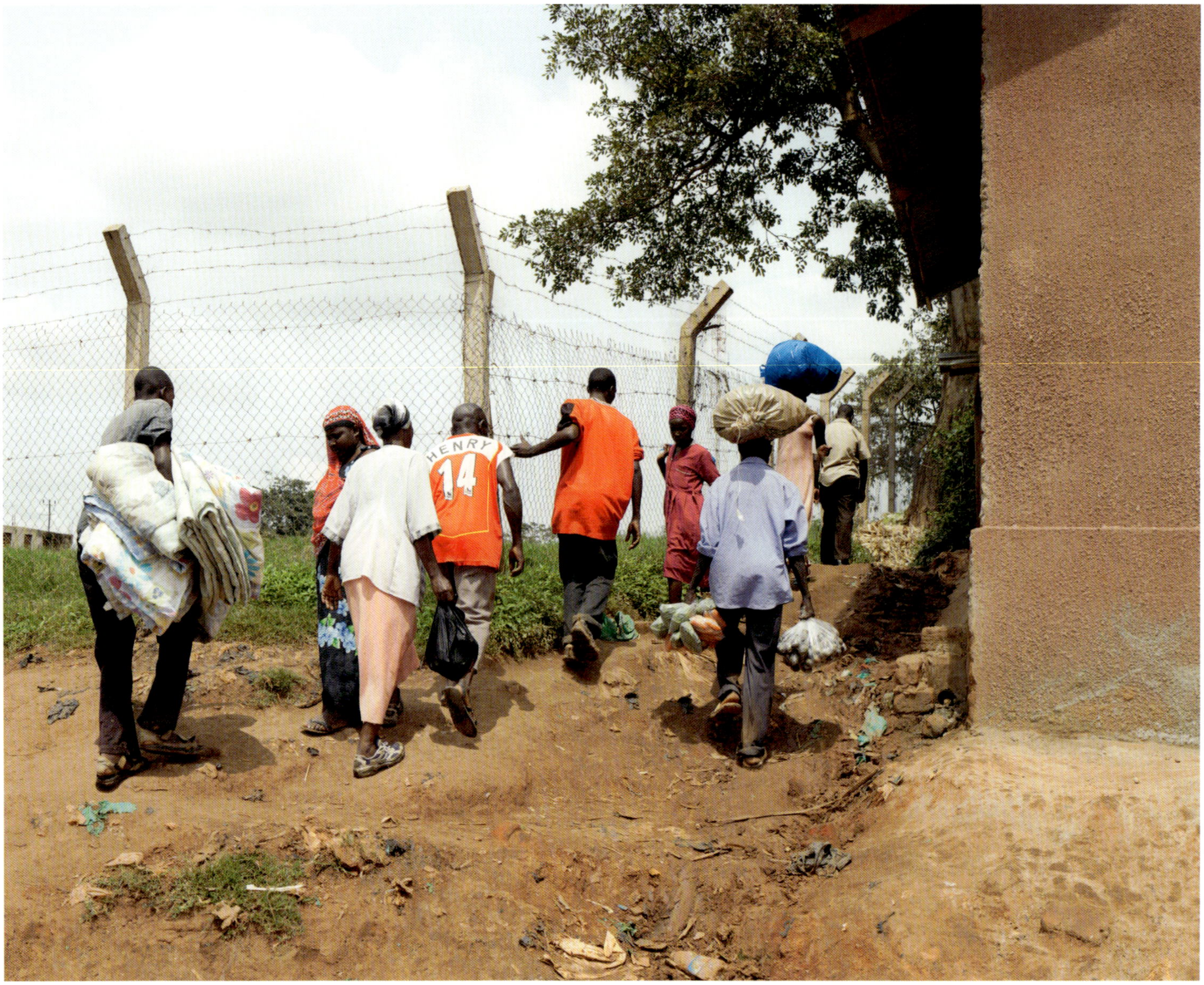

BWAISE (PATH)
KAMPALA

Recognizing and Revealing the Public

ALEXANDER OPPER

The term "photography" does not suffice to describe the work of Lard Buurman. The twenty-first century, for all its advances and overload in terms of the availability and retrievability of information, is severely behind regarding the necessary (re)definition of descriptors for hybrid artistic practices, of which Buurman's systematic photographic constructions are a noteworthy example.

PUBLIC SPACE

In the context of reflecting on Lard Buurman's work as a "record" of public space I read his role, on one level, as that of a photographer-builder. His working method stems from an approach of collecting site-specific "material." He literally (digitally) collects passersby as they traverse and pause on their respective everyday journeys across his camera's selective field. All these figures inadvertently form part of a casting of which only Buurman is aware: some of these urban actors will make it onto the flat surface of one of his constructed stage-set-like tableaus, some will not.

Buurman, via the constructed nature of his composite photographic images, avoids the risky lure of the single shot to depict or, even bolder, attempt to *represent* a truth to context. His work adds to a discourse, which over the past decade or two, has slowly begun to grapple with long overdue new readings of Henri Lefebvre's concept of the "production of space." I see Buurman's work as a valid attempt at interpreting and visually translating the *co*-production of public space. His images expose the value of the essentially ungraspable, namely the informal or spontaneous nature of how the public is produced in many African urban and rapidly urbanizing centers. Spontaneity and informality should not be misread here as romantic smokescreens for a layer of precariousness not immediately apparent in this collection of images; nor as syntactical distractions from the multiple livelihoods practiced by the anonymous figures populating these fabricated scenes. Mobility and endurance are the currencies that connect his images to a multiplicity of *other* locations of everyday lived practices. If we look closely at some of the images, the uncanny Muybridge-like reappearance of actors in a single frame hints at trajectories of necessity; the necessity for multiple ways of being and enduring in order to limit, as much as possible, the reliable presence of precariousness.

PEOPLE AS INFRASTRUCTURE

What Buurman's images do is to shift our focus from a dumbed-down desire for Western-premised typological spatial foci (such as "public" squares, "public" parks, and so forth)—as easily measurable and categorical urban building blocks—to careful observations centered on "people as infrastructure." This idea, after AbdouMaliq Simone's essay (2004) of the same name, questions a still pervasive subscription to entrenched urban design conventions. It challenges the unquestioned significance often attached to urban infrastructural *hardware* rolled out in the form of decisively designated "public" spaces. Simone's observations redirect attention from the typically "hard" focus to the infrastructural importance of the flux-based human bodies of individuals and groups. Seen from this perspective, traces of permanently shifting bodily placements and the transactions between these begin to crystallize in the frozen condition of Buurman's images. They start to uncover how, in a relational and almost impossible-to-measure rhizomic sense, the co-production of space—and sometimes *place*—is based on a range of complex individual and networked motivations and negotiations. Linked to this reading of the centrality of the body in the activation of the public realm, Buurman's compositions help to expose the narrow limits of outdated and oversimplified Western conceptions. Namely, of neatly quantifiable notions of togetherness or communality—as epitomized by, for instance, uncritical categorical and typological considerations of "parks" and "squares" as being universally applicable, and successful as spaces of inclusion and belonging.

It should be understood that the above distinction, which focuses on the integral value of people in terms of the generation of public space and place, does not deny the crucial nature and necessity, for any city, of more traditional conceptions of infrastructure, vis-à-vis the provision of formal things such as stable roads, sanitation, and electricity for citizens. What Buurman's choreographed images point to is a reinterpretation of the so-called formal. They show that what we perceive of as fixed formal structures and the informal ways in which users appropriate the formal—opportunistically—are two legitimate systems that riff off each other.

Buurman's collaged images expose how the formal traces—the architectural and urban residue of the colonial pasts of the cities photographed—have been productively adopted as not simply a backdrop, but a malleable support: a framework for public expression and public futures. His images reveal radically optimistic manifestations of instances of a bottom-up, people-driven attitude to the co-production of public space. Together, the images talk about a *non-figurative* and fluid interpretation and translation of previously more mono-functionally conceived and prescriptive spaces. The "non-figurative" register I refer to is meant to draw attention to the need for a dynamic interrogation of the modernist use of the figurative description of a functionally-ascribed "body" of space as either a "positive" space (i.e. a building) or, its corollary, a "negative" space (i.e. open spaces—such as the street, square, and park typologies). This photographic revelation of a grassroots taking-ownership-of, by the urban actors of the everyday, challenges traditionally top-down and institutionally-driven preconceptions of more normative and invariably Western-defined understandings of public space, what it looks like, and how it is imagined to be "correctly" used. What is happening in these images is not the tame "appropriation" of space, which has become so fashionable a descriptor, but the *possession* of space. Such spaces become use-full and filled with use—they are produced public spaces that are simultaneously productive. Most importantly, they are alive.

THE STREET

Buurman's pictures begin to reveal to us a lived imagery of what public space actually *is*, as opposed to what it "could" or "should" be. All the actors who make appearances on the final images actually physically crossed the captured scenes. He respectfully collapses the "life worlds" of multiple urban participants into a "contact zone" of sorts. Here I borrow from James Clifford's (1997) use of this term, coined in order to challenge—in a different context—the relevance of highly selective and edited museum environments. The transferal of the contact-zone attitude seems to me appropriate as a prompt for the recognition and the naming of the rich range of both planned and more spontaneous occurrences that play out along a particularly crucial building block for any city, namely the street. The urban spine of the street contains the DNA that allows it the potential, through use, to translate into a universal equalizer. It is not vested with the narrower programmatic limits of the urban square or the park. In its immediate and obvious functional guise of connectivity and connectedness it serves simultaneously as a perfectly neutral backdrop for, and loaded vessel of, urban life.

It is in this rich urban constituent, the *unbuilt* zone of the street, where the building of public life is achieved. The street I am talking about here is neither related to the squeaky-clean finality of a Zurich body "public," nor to the imposed limitations attached to the privatized public space of Berlin's Potsdamer Platz. Although the privatization of public space has made its presence felt within speculative aspects of city making (or should that read city-*unmaking*?), in larger cities on the African continent, Buurman's framing of the street-life condition speaks of this equalizer as a connective and relatively free scape of engagement, surprise, deal-making, and opportunity. His assemblages clearly allude to the notion of the fluidity of "people as infrastructure." His images reveal how streets, by nature of their position and what they connect to, offer canny users possibilities. Possibilities for the superimposition of value and opportunity—socially and economically—that question the definitive, often absurd urban rules with which we are all familiar. Namely, the "Do's and Don'ts" we have all encountered at entrances to more traditional "public" parks and squares. The streetscape, which forms the stage for many of the constructed images in this book, demonstrates the power of the street as the embodiment of non-building: a space that allows for the translation and transformation of desires and encourages true publicness—as an opportune and productive merging of a diversity of entangled practices of everyday life.

ARCHITECTURE

Buurman fills his photographic portrayals of streets to the brim, more or less. His method invariably pushes the architectural and urban setting to the status of an extra. This is refreshing, when compared to the disembodied and ever-popular photographic representations of architecture celebrated in the media: images devoid of the human body, thus denying its attendant potential to produce and (re)produce public space. Architecture's reductive avoidance of the abject body of the human user (better said, in the case of so many modernist failures, the victim) has been challenged by key exponents of a practice of counter-modernism. As long ago as the nineteen-fifties and sixties, the architect couple Alison and Peter Smithson, practitioners, theorists, and teachers, reminded us of the value of the street, along with all its messy attachments.

A relatively recent and absolutely necessary revisiting of the importance of the *lived* practice of everyday life, as expounded by the Smithsons and others has taken place. This thanks largely to a wake-up call, to an insular architectural profession, to self-assess and recognize the importance of keeping an ear to the street, so to speak; with valuable lessons to be learned from the intelligences exhibited through the informal production and use of space, in the so-called "third" or "developing" world. Despite this reawakening within part of the architectural fraternity the selling of the "clean" architectural image still clings on for dear life, quite successfully I might add, still one-sidedly influencing the more privileged popular imagination as to what architecture "is."

Architectural stardom is so inherently linked to such pared-down representations of the power of (inanimate and unanimated) architecture, that the works of Buurman become important reminders that reinforce the urgency of the wake-up call alluded to above; a reminder that celebrates the emergent and hard-to-define breadth and richness of the multiple expressions of truly public life on the streets of so many African cities. These photographs talk about a value that is tenuously linked to notions of freedom (to be, to use, to express . . .) that is by no means simple or, for that matter, guaranteed; an emergent grappling with a freedom that was, until recently, severely curtailed by the political, urban, and architectural

apparatuses of enforced colonial rule. These images express a tentative current experimentation of what freedom and a "right to the city" might mean for citizens belonging to postcolonial urban contexts which, in many cases, suffer from newer forms of postindependence and state-driven restriction of the degrees of permitted publicness.
A good example of these current strictures is well demonstrated by the City of Johannesburg's recent so-called clean-sweep of informal traders trading in Johannesburg's Central Business District. This top-down decision has resulted in the violent erasure of a layer of self-organized, livelihood-dependent exchange that guaranteed, until recently, a palpable lived production of street-driven "cityness," synonymous with and integral to the inner-city's aura and dynamic people-driven identity. The precariousness of this Johannesburg story can be found, in many other forms in the multitude established and rapidly growing urban centers across the continent. The meaningful emergence of manifestations of publicness in the complex urban contexts portrayed in this book is invariably exposed to hidden power relationships and agendas, which cannot be conveyed on the flattened surfaces of these photographs. The places we see hold equal portions of precariousness and possibility for their users. Buurman's constructed images avoid a naïve belief that emergent notions of inclusive spaces of participation and belonging will be easy for citizens to achieve and hold on to.

To conclude this brief reflection on the difficult task of the recognition and revelation of publicness in the multifaceted loci of African urbanism, I believe it is important to acknowledge the value of how Buurman pushes and tugs at the generally static determinants of his discipline, photography. Buurman's photographic constructions begin to tease out the uncannily subtle possibilities of disciplinary crossings and productive manipulations. His attempts at provocative translations of almost-impossible-to-capture everydayness become particularly significant in the light of how we might begin to appreciate the complex and necessarily flexible meanings of the staging of public life in the contemporary African city. Just as the subjects in his images occupy and make the streets public, so Buurman occupies and crowds his images. Via the methodical postproductive population of his accretive composite images, he achieves something similar to a congestion of flattened Godard-like jump cuts. His working process allows for the cojoining of the static instantaneity of the single photographic snap with the energy inherent in the motion upon which the younger discipline of cinema relies.

Hybridity seems an apposite register for Buurman's constructed "documentary" practice. I see this bridging and hybridizing of different disciplinary traits as an appropriate attempt, not at a *capturing*—in a simplistic sense, which assumes an understanding of the complex codes and layers of publicness Buurman addresses in his work—but as a *relaying* of aspects or hints at what "public" might mean in the multifarious contexts of the urban centers he has visited, recorded, and ultimately constructed, on the African continent over the past few years.

List of Works

Author Biographies

Chris Abani (*Nigeria, 1966) is an acclaimed author. His most recent novel is *The Secret History of Las Vegas*. He is the recipient of a Guggenheim Fellowship, the PEN/Hemingway Award, the PEN Beyond Margins Award, the Hurston Wright Legacy Award, and a Lannan Literary Fellowship, among many other honors. Born in Nigeria, he is currently a Board of Trustees Professor of English at Northwestern University. He lives in Chicago.

N'Goné Fall (*Dakar, 1972) graduated from the École Spéciale d'Architecture in Paris. She is an independent curator, an essayist, and a consultant in cultural policies. She served as the editorial director of the Paris-based contemporary African art magazine *Revue Noire* from 1994 to 2001. She has edited books on contemporary visual arts and photography in Africa including *An Anthology of African Art: The Twentieth Century, Photographers from Kinshasa* and *Anthology of African and Indian Ocean Photography: A Century of African Photographers*. Fall has curated exhibitions in Africa, Europe, and the United States. She was one of the curators of the African Photography Biennial in Bamako in 2001 and a guest curator of the 2002 Dakar Biennial. As a consultant in cultural policies she is the author of strategic plans, orientation programs, and evaluation reports for Senegalese and international cultural institutions. Fall is an associate professor in the Department of Cultural Industries at the Senghor University in Alexandria, Egypt. She is also a founding member of the Dakar-based collective GawLab, a platform for research and production in the field of technology applied to artistic creativity.

Chris Keulemans (*Tunis, 1960) is the artistic leader of Tolhuistuin, the new arts center in Amsterdam Noord. As a writer and journalist, he has published six books of fiction and non-fiction, and numerous articles on artists such as Fiona Tan, Jonas Staal, Tony Chakar, and Douglas Gordon to name a few. He founded a website on art after crisis, investigating how art is revitalizing cities like Tirana, Beirut, New Orleans, Sarajevo, Algiers, and Jakarta.

Alexander Opper (*South Africa, 1972) lives and works in Johannesburg, as an educator, writer, artist, architect, and designer. He is the senior lecturer and director of the MTech Architectural Technology program in the University of Johannesburg's Faculty of Art, Design, and Architecture department. Here he practices a teaching-and-learning approach premised on "folding" the architectural studio into the field. As an artist, his site-specific interventions subvert entrenched codes attached to the white cube and interrogate uneven power relationships embedded in, and in-between, urban conditions. His artistic output and his writings rely on an approach he refers to as "undoing architecture." In 2013 Opper's debut solo exhibition—*Separ(n)ation*—completed its run at GoetheonMain, Johannesburg.

Acknowledgments

I would like to express my sincere gratitude to those mentioned in the list below, who have kindly and competently assisted or advised me over the past six years. Their contributions have been indispensable to this project.
Special thanks are also due to my assistants throughout the African continent, who have guided me in finding my way in their cities and who proved to be of vital importance throughout the course of my project.
In addition, I would like to thank all those people who I have encountered in my journeys and of whom I have a specific memory, but no name.

CAIRO
Devon Cone, Ahmed el Ghamrawi, Gianluigi Guercia, Dina Kafafi, Khaled (Saqqara), Moataz Nasr, Philip Rizk, Dirk Wanrooij

CAPE TOWN
Tau Tavengwa (African Centre for Cities)

CASABLANCA
Florence and Hassan Darsi (La Source du Lion), Hicham Abou Lmajd, Rashid Kinoj, Mohamed el Khadiri

DAKAR
Karen Dermineur, Thio Dia Ibrahima, Jean-Charles Tall, Pascal Nampémanla Traore, Mama Wane and Nadia Rodriguez (Timtimol)

DAR ES SALAAM
Hein Bouwmeester, Nikolas Mombo

DOUALA
Marilyn Douala-Bell, Doual'Art, Xandra Nibbeling, Kamiel Verschuren

JOHANNESBURG
Lester Adams, Laurence Bonvin, J. J. de Castro Maia, Jordache Ellapen, Raymond Marlowe, Maria Fidel Regueros, Gabriël Frankel and Jonathan Liebmann (Hotel 12 Decades), James French (Bag Factory), Bronwyn Lace and Marcus Neustetter (LiveCollection), Phillip Legate, Charles Ndumo, Alexander Opper, Chloe Sweerts, Myer Taub

KAMPALA
Mister Bosco, Jeroen de Lange, Marcia Luijten, Margaret Nagawa

KINSHASA
Henri Kalama Akulez, Sammy Baloji, Vitshois Mwilambwe Bondo, Julie Djikey, Meike de Goede, Marco Hekkens, Tshoper Kabambi, Kiripi Katembo, Sylvain Nzuzi Nsaia, Mama Paula

LAGOS
Omon Anenih-Mordi, Isi Etomi, Doyin Hassan, Azu Nwagbogu (African Artists' Foundation), Emeka Mordi, Giles Omezi (London), Tsunde

LUANDA
Job Beeckmans en Machteld Cattrysse, Edson Chagas, Yolande Ditewig, Hotel Globo, Carlos Major, José Manuel (Mano), Marita Silva (Fundaçao Sindika Dokolo), Mulumba Seraphin, Eleni Thomaidis, Trienal de Luanda

LUSAKA
Kaleb de Groot, Baaba Jakeh, Ngandwe Mwaba, Mabvuto Mwanza, Rockston Studios

NAIROBI
Adigo Digo, Fleur van Dissel, Duncan, Kevin Iningu, Jungle Junction, Sidney Mongongo

NETHERLANDS
Kévin Bray, Bonnie Dumanaw, Lino Hellings, Juul Hondius, Joris Jansen, Pjotr de Jong, Mayke Jongsma (Mondriaan Fund), Hans Kennepohl, Berend van der Lans, Viviane Sassen, Wim van Sinderen, Frank van der Stok, Steven van Teeseling (Mondriaan Fund), Hugo Timmermans, Astrid Vorstermans, Eric Wie, Hans Wilschut

SPECIAL EDITIONS
Willem Asselbergs (Mertens Frames, Amsterdam), André Beuving (BLXM), Gertjan and Bibet Buurman-Wentink, Kai van Hasselt, Huub Duijsens, Marco Hekkens, Bjarne Mastenbroek (SeArch, Amsterdam), David Snellenberg (Dawn, Amsterdam)

For this project the following persons and their work have been inspirational to me:
Filip de Boeck, AbdouMaliq Simone, Jordache A. Ellapen, Achille Mbembe, Garth A. Myers, Edgar Pieterse, Omar Ngati

Colophon

Editor: Nina Folkersma
Copyediting: Leina González
Translation: Adre Marshall (Fall essay)
Image editing: Lard Buurman and Roosje Klap
Graphic design: Roosje Klap
Typefaces: Atlas Grotesk, Publico Banner, Ano
Production: Ines Sutter, Hatje Cantz
Reproductions: Sebastiaan Hanekroot, Colour&Books
Paper: Schleipen Fly bläulich weiss, 90 g/m²;
Novatech Ultimatt 1.1, 135 g/m²
Printing: Koninklijke Van Gorcum, Assen
Binding: Boekbinderij P. Vh Abbringh, Groningen

Published by
Hatje Cantz Verlag
Zeppelinstrasse 32
73760 Ostfildern
Germany
Tel. +49 711 4405-200
Fax +49 711 4405-220
www.hatjecantz.com
A Ganske Publishing Group company

Hatje Cantz books are available internationally at selected bookstores. For more information about our distribution partners, please visit our website at www.hatjecantz.com.

ISBN 978-3-7757-3791-3
Printed in the Netherlands

This project and publication have been made possible through the generous support of the Mondriaan Fund

M
mondriaan
fund

Cover illustration:
Ponte de Villa du Gamek, Luanda, Angola
(2009 / 2010)

Back cover illustration:
Setswetla, Alexandra, Johannesburg, South Africa
(2009 / 2011)